Fertilization

Outline Studies in Biology

Editors' foreword

The student of biological science in his or her final years as an undergraduate and first years as a graduate is expected to gain some familiarity with current research at the frontiers of his or her discipline. New research work is published in a perplexing diversity of publications and is inevitably concerned with the minutiae of the subject. The sheer number of research journals and papers also causes confusion and difficulties of assimilation. Review articles usually presuppose a background knowledge of the field and are inevitably rather restricted in scope. There is thus a need for short but authoritative introductions to those areas of modern biological research which are either not dealt with in standard introductory textbooks or are not dealt with in sufficient detail to enable the student to go on from them to read scholarly reviews with profit. This series of books is designed to satisfy this need. The authors have been asked to produce a brief outline of their subject assuming that their readers will have read and remembered much of the standard introductory textbook of biology. The outline then sets out to provide, by building on this basis, the conceptual framework within which modern research work is progressing and aims to give the reader an indication of the problems, both conceptual and practical, which must be overcome if progress is to be maintained. We hope that students will go on to read the more detailed reviews and articles to which reference is made with a greater insight and understanding of how they fit into the overall scheme of modern research effort and may thus be helped to choose where to make their own contribution to this effort. These books are guidebooks, not textbooks. Modern research pays scant regard for the academic divisions into which biological teaching and introductory textbooks must, to a certain extent, be divided. We have thus concentrated in this series on providing guides to those areas which fall between, or which involve, several different academic disciplines. It is here that the gap between the textbook and the research paper is widest and where the need for guidance is greatest. In so doing we hope to have extended or supplemented but not supplanted main texts, and to have given students assistance in seeing how modern biological research is progressing, while at the same time providing a foundation for self-help in the achievement of successful examination results.

General Editors:

W.J. Brammar, Professor of Biochemistry,
University of Leicester, UK

M. Edidin, Professor of Biology,
Johns Hopkins University, Baltimore, USA

Fertilization

Frank J. Longo
Department of Anatomy
The University of Iowa, USA

LONDON   NEW YORK
Chapman and Hall

First published in 1987 by Chapman and Hall Ltd
11 New Fetter Lane, London EC4P 4EE
Published in the USA by Chapman and Hall
29 West 35th Street, New York NY 10001
© 1987 F.J. Longo

Printed in Great Britain at the
University Press, Cambridge

ISBN 0 412 26880 9 (Hardback)
0 412 26410 2 (Paperback)

British Library Cataloguing in Publication Data

Longo, Frank J.
  Fertilization.
  1. Fertilization (Biology).
  I. Title
  591.3'33   QP273

  ISBN 0–412–26880–9
  ISBN 0–412–26410–2 Pbk

Library of Congress Cataloging in Publication Data

Longo, Frank J., 1939–
  Fertilization.

  (Outline studies in biology)
  Includes bibliographies and index.
  1. Fertilization (Biology) I. Title. II. Series:
  Outline studies in biology (Chapman and Hall)
  [DNLM: 1. Fertilization. 2. Germ Cells–physiology.
  QH 485 L856f]
  QP273.L66 1987   591.3'3   87–15075
  ISBN 0–412–26880–9
  ISBN 0–412–26410–2 (pbk.)

Contents

Preface

E.B. Wilson stated in the preface to the third edition of *The Cell in Development and Heredity*: 'Every writer must treat the subject from the standpoint given by those fields of work in which he is most at home; and at best he can only try to indicate a few of the points of contact between those fields and others'. The aim of this book is to provide an overview of structural and functional aspects of fertilization processes in a manner that would be helpful not only to specialists in the field but also to investigators in related disciplines and to advanced undergraduate and graduate students in the biological sciences. Fundamental descriptive accounts at the light and electron microscopic levels of observation have been combined with analytical studies – physiological and biochemical investigations of fertilization in invertebrates and vertebrates. A comparative approach to fertilization is presented and, although a variety of animals are referred to, additional space is purposely given to organisms that have been, and continue to be, popular research material.

The text does not pretend to be comprehensive and admittedly does not cover all aspects of the field or areas related to the general subject. Historical reviews and technical details are presented only to the extent necessary to formulate an orientation to and a perspective on individual topics. Areas such as fertilization mechanisms in plants, gamete development and the role of accessory reproductive structures on fertilization, particularly in mammals, have been included only to a limited extent in an effort to keep the book within the bounds of an overview.

It is not possible to list or refer to even a major portion of the vast literature on fertilization. In this book, for the most part, each topic is referenced with reviews and/or original research articles that pertain to several different organisms and include earlier and more contemporary investigations of the subject. It is anticipated that this approach may aid the reader in exploring the depth and nuances of a particular topic.

In any venture such as this credit is due to a number of individuals who have been generous with their time and efforts. Because my own views of fertilization processes come in large measure from previous mentors, I acknowledge their presence in these pages, particularly Everett Anderson. Sincere thanks and grateful appreciation is expressed to Becky Hurt, Tena Perry and Vicki Fagen for typing the manuscript; to Frederic So and Shirley Luttmer for their assistance in the organization of the references; to Julie Longo, Brenda Robinson and Paul Reimann for their artistic help in the preparation of the illustrations, and to Jo Ann Barnes for her generous and conscientious editorial assistance. Kitty, Julie, Joe, Crista, Thad, Jude, Gabrielle and Gian Carlo are also owed special thanks for many special favors.

1

General considerations of fertilization: A definition

The interaction of the spermatozoon and the egg initiates a series of transformations involving the nuclear and cytoplasmic components of both gametes. These transformations constitute the process of fertilization, which has been defined as a multistep phenomenon commencing with the interaction and subsequent fusion of the gametes and culminating in the association of the corresponding groups of chromosomes derived from two pronuclei, one of maternal and the other of paternal origin (Wilson, 1925). In almost all cases investigators have pointed out that the essential aspects of fertilization are: (a) the association of the maternal and paternal genomes – biparental heredity, and (b) the activation of both the sperm and the egg – a series of events which alters the metabolism of both gametes and leads to the cleavage and differentiation of the fertilized egg or zygote.

Activation of the egg can be initiated by processes other than fertilization. Parthenogenetic activation by chemical and/or physical stimuli may lead to complete development or to the initiation of processes that simulate fertilization but do not give rise to viable offspring. In either case, these observations indicate that the egg is endowed with the essential machinery and information to initiate developmental processes when suitably stimulated. Furthermore, activation changes that occur at fertilization and parthenogenesis do not require immediate gene action, implying that the genetic activity required for establishing the fertilization response occurs during oocyte maturation. Metabolic changes in the egg evoked by fertilization ultimately affect new gene expression and differentiation during later stages of embryogenesis. Hence, fertilization in the scheme of a biparental organism serves as a point of transition between gamete and embryonic development.

Investigations of fertilization date well before the turn of the century under the leadership of cytologists such as Van Beneden, Flemming, Strasburger, Boveri, and Wilson whose observations on germ cells were closely affiliated with theoretical writings of Nageli, Weissmann, Hertwig, Roux, and De Vries (Wilson, 1925). Their remarkable observations and formulations provided the intellectual framework for the chromosome theory of inheritance. More contemporary research on fertilization tends to serve a dual role, one being its application to fertility control, and the other as a 'model system' to study basic processes of cells in general.

Much of the research in fertilization has employed the gametes of invertebrates, particularly the sea urchin. The reasons for the popularity of this animal include its availability and minimal requirements for maintenance. In addition, because the sexes are separate and large quantities of gametes can be obtained, which can be fertilized externally and develop in synchrony, one may analyze processes using a wide variety of techniques. Consequently, important insights have been gained into the causal chain of events that occur during fertilization in this animal, which are relevant to the study of fertilization in higher organisms. These insights reveal the nature of the changes that accompany and follow the interaction and fusion of the gametes. The timing of many of the events comprising fertilization in sea urchins is given in Table 2.1, p.4. Studies using the gametes of invertebrates and vertebrates indicate that a comparable sequence of events is initiated in both groups of animals; sperm—egg attachment leads to the 'turning on' of new synthetic activities and developmental programs. Specific aspects of such studies have been recently reviewed (Metz and Monroy, 1985).

Since metabolic and morphological changes of sea urchin and mammalian gametes during fertilization have been well characterized experimentally, studies employing these organisms provide the major thrust of what is discussed herein. Important data, obtained from the study of eggs from organisms such as molluscs, annelids, tunicates, fish and amphibians are also presented. The two principal events of fertilization considered are: (a) the initial interaction and activation of the sperm and egg, and (b) the concluding events, involving pronuclear development and association, that eventually lead to cleavage.

2

Sperm activation

Motility and behavioral changes

As a rule sperm are stored inactive until fertilization. This dormancy is partly maintained in sea urchins by a low intracellular pH (Lee, Johnson and Epel, 1983). Sperm activation at fertilization is initiated by a series of changes in ion content resulting in an increase in internal pH. There is a sodium-dependent acid release coincident with an increase in sperm motility upon the dilution of semen into sea water. The activating factor does not appear to be sodium *per se* as incubation of sperm in solutions of high external pH (pH 9) or NH_4Cl, which presumably increase the intracellular pH, also activate motility in the absence of sodium.

Egg jelly preparations from some invertebrates stimulate sperm respiration and motility. When sea urchin sperm are incubated in sea water at an acid pH they are immotile and have a low respiration rate. However, if egg jelly is added respiration and motility increase dramatically (Hansbrough and Garbers, 1981a,b). The active agent of sea urchin egg jelly has been shown to be a glycine-rich polypeptide of approximately 2000 molecular weight. This material (speract) stimulates respiration and motility in a non-species specific fashion in sea urchins, but does not induce the acrosome reaction. It is most potent in stimulating sperm at acidic pH values, increasing respiration rates by as much as tenfold.

At pH 6.6, in the presence of speract, sperm undergo a sodium influx and a sodium-dependent proton efflux (Hansbrough and Garbers, 1981b). Proton release appears to be responsible for motility initiation and the increase in respiration, since both can be induced in sea water at pH 6.6 by monensin, a sodium/proton ionophore. If sodium is not present the increase in respiration fails to occur. Activation of respiration appears to be potassium and calcium independent.

Coincident with an increase in respiration and motility, speract also brings about an increase in sperm cAMP and cGMP (Hansbrough and

Table 2.1   Timing of fertilization events in such urchin eggs. Timing of these events depends markedly upon species and temperature. Values given are for *Lytechinus pictus* at 16–18°C. See Whitaker and Steinhardt (1982).

Membrane potential	
Ca–Na action potential	Before 3 s
Na activation potential	3–120 s
K conductance increase	500–3000 s
Intracellular calcium release	40–120 s
Cortical granule reaction	40–100 s
NAD kinase activation	40–120 s
Reduced nicotinamide nucleotides increase	40–900 s
Acid efflux	1–5 min
Intracellular pH increase	1–5 min
Oxygen consumption increase	1–3 min
Protein synthesis initiation	5 min onwards
Amino acid transport activation	15 min onwards
DNA synthesis initiation	20–40 min

Garbers, 1981a,b). Observations demonstrating that similar changes are elicited by bromo-cGMP and that analogs of cAMP are without effect, suggest that cGMP may mediate, or be closely associated with, the increases in sperm respiration and motility induced by speract.

In addition to increases in motility and respiration, when sea urchin sperm are incubated in solubilized egg jelly they undergo an agglutination which has been referred to as swarming (Lopo, 1983; Fig. 2.1). Similar phenomena have been described for the sperm of other organisms. The affected sperm aggregate into dense clusters 2–4 mm in diameter and 5–10 min later they spontaneously disperse. The dispersed sperm do not swarm again with the addition of fresh jelly. The swarming of sperm when exposed to egg jelly provided the basis for the fertilizin–antifertilizin theory of Lillie (1919). This 'isoagglutination' reaction was thought to represent a reversible complex formation comparable to the interaction of antibody with antigen. Reversibility was brought about by a modification of agglutinating substances of the egg jelly (fertilizin) by sperm from a multivalent to a univalent form (Metz, 1967). More recent investigations, however, indicate that this process is not a true agglutination as the sperm do not contact one another. Furthermore, the process is reversibly blocked by respiration inhibitors indicating that it is dependent upon sperm motility (Collins, 1976).

The possible role of sperm swarming during fertilization has not been clearly defined; in fact, it does not appear to be obligatory since successful fertilization can occur in its absence. Interpretation of its

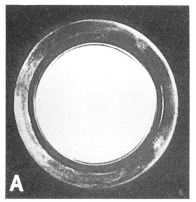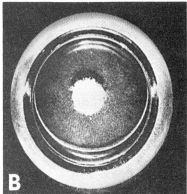

Figure 2.1 Agglutination of *Megathuria* (mollusc) sperm induced by egg water. (A) Dense sperm suspension before addition of egg water; (B) aggregated sperm 10 min after the addition of egg water. Reproduced with permission from Tyler (1940).

possible significance at fertilization is paradoxical. It may function to keep excess sperm from reaching the egg and, ultimately, act to reduce the concentration of sperm in the immediate vicinity of the ovum thereby preventing polyspermy. Conversely, it may attract sufficient numbers of sperm to the vicinity of the egg, thereby promoting successful fertilization.

During epididymal transit mammalian sperm acquire the capacity for motility. This involves a transition from a circular- or whiplash-type of movement of the caput sperm to a vigorous forward motion characteristic of sperm from the cauda (Hamilton, 1977; Hoskins, Brandt and Acott, 1978). This change does not appear to depend upon the available energy supply. Cyclic nucleotide phosphodiesterase inhibitors markedly increase respiration (three-to fourfold) and motility of cauda epididymal sperm. The observed stimulating effects on motility are mediated by an increase in intracellular cAMP (Garbers, First and Lardy, 1973; Garbers *et al.*, 1973).

The progressive motility acquired by mammalian spermatozoa during their maturation within the epididymus involves: (a) an increase in intracellular cAMP, and (b) the binding of a specific component, forward motility protein, to the sperm surface (Hoskins *et al.*, 1979). Forward motility protein has been characterized as a heat stable glycoprotein of 37 500 molecular weight that originates within the epididymus. Its activity has also been found in seminal plasma.

The effect of cAMP on mammalian sperm motility may be direct or mediated by protein phosphorylation. Mammalian sperm have a cAMP-dependent protein kinase and phosphorylation of a 55 000 molecular weight protein is correlated with sperm motility, suggesting that the effect of cAMP upon sperm is mediated via protein phosphorylation (Garbers, First and Lardy 1973; Garbers *et al.*, 1973). Major proteins of the sperm tail, e.g. tubulin and dynein, are among sperm polypeptides phosphorylated in association with cAMP enhanced motility.

Before the acrosome reaction, hamster sperm move vigorously, they form figure eights by a whiplash-like beating of the flagellum (Fig. 2.2). This change in sperm motility has been termed hyperactivation (Yanagimachi, 1981). Whether hyperactivation is a general phenomenon, common to all mammals, has not been determined. Moreover, the physiological significance of this change in activity is unclear. Nevertheless, activated motility is also seen in rabbit sperm and is characterized *in vitro* by episodes of non-progressive swimming with large amplitude whiplash-like flagellar undulations, alternating with low amplitude progressive swimming. A similar type of movement is associated with hamster sperm taken from the ampulla at about the time of fertilization. This increase in activity may represent a means whereby the sperm can exert strong thrusting movements during transit through layers surrounding the egg, the cumulus and the zona pellucida.

Alterations in the distribution of intramembranous particles occur within the plasma membrane of the sperm middle piece at the time of hyperactivation and may be related to this change in sperm motility (Koehler and Gaddum-Rosse, 1975; Friend *et al.*, 1977). Epididymal and ejaculated hamster sperm treated with Triton X-100 and incubated in buffer containing ATP are able to exhibit hyperactivation, suggesting that the flagellum is prevented by some mechanism from affecting the activated type of movement.

Chemotaxis

One of the major pitfalls of early studies examining the possibility of sperm chemotaxis in different organisms was the trapping of sperm in extraneous layers surrounding the egg. Results of these studies gave the impression that sperm were attracted to the ovum. Experiments examining turning movements of spermatozoa to sources of putative attractants have demonstrated such a process in a number of animals and it is possible that sperm chemotaxis may be more widespread than previously realized (Miller, 1977). Chemotaxis has been demonstrated in species of coelenterates, tunicates and chitons, where sperm show a

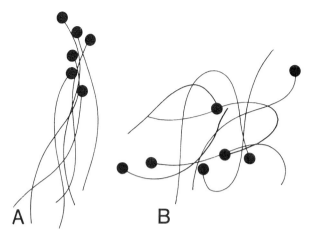

Figure 2.2 Shapes of typical flagellar movements of epididymal (A) and hyperactivated (B) hamster spermatozoa. See Yanagimachi (1981).

gradient directed turning behavior, i.e., they reorient and swim towards a putative attractant.

The fact that chemotaxis has been demonstrated in only a few unrelated species suggests that this phenomenon may have evolved independently on several occasions and that it may be a prerequisite to successful gamete interaction. The possible advantage of such a process is that release of an attractant would increase the probability that sperm will reach the ovum. In a sense, it effectively increases the volume of the egg as a target for the spermatozoon.

The capsule of the siphonophore, *Muggiaa*, has been shown to attract sperm (Sardet *et al.*, 1982; Fig. 2.3). Sperm approaching capsule material reduce the curvature of their trajectories from slightly curved paths to small circles so that eventually a cloud of sperm forms around the attractant source. During this attractive phase sperm flagellar beat frequency and velocity are not noticeably modified. The active substance isolated from capsules has an apparent molecular weight of about 67 000. Calcium is required for sperm to respond to the attractant. This calcium dependency can be mimicked by treating sperm with the calcium ionophore A-23187, suggesting that the attractant induces a calcium flux, thereby increasing the asymmetry of the flagellar beat pattern.

What turns off the chemotactic stimulus has not been determined. Sperm chemotaxis disappears following fertilization or artificial activation possibly as a result of cortical granule exocytosis, although the

Figure 2.3   Accumulation of sperm at the animal pole of a *Muggiaea* egg. See Carre and Sardet (1981).

involvement of other egg changes have not been eliminated.

Capacitation

As the mammalian spermatozoon passes through the epididymus it undergoes changes which ultimately give rise to its mature form. The cytoplasmic droplet migrates caudally along the middle piece and is eventually lost. There is an increase in the stability of the nucleus, basal plate, connecting piece, outer dense fibers, fibrous sheath and mitochondria, which has been attributed to an increase in intra- and intermolecular cross-links (Calvin and Bedford, 1971).

During epididymal maturation the spermatozoon acquires the capacity for progressive movement (Hamilton, 1977; Hoskins, Brandt and Acott, 1978). Although this change involves components of the axonemal complex, it is believed to be regulated by alterations associated with the plasma membrane. The distribution of lectin binding sites on the sperm surface may increase, decrease or remain unchanged – depending upon the species, functional domain and the protocol employed (Nicolson and Yanagimachi, 1979). Surface labeling studies using ^{3}H-sodium borohydrate or lactoperoxidase catalyzed iodination have been used to demonstrate differences in membrane proteins from caput and cauda sperm (Olson and Danzo, 1981). Immunochemical probes have shown a loss and/or modification of glycoproteins during epididymal maturation of mouse and boar sperm (Feuchter, Vernon and Eddy, 1981). In addition to modifications of intrinsic sperm plasma membrane proteins, glycoproteins are secreted by cells of the male accessory reproductive organs and become associated with the sperm plasma membrane. The association of these components with the spermatozoon may account for surface changes such as charge density and adhesiveness.

Mammalian sperm require an additional phase of maturation, occurring within the female reproductive tract, to prepare them for the acrosomal reaction and fertilization (Yanagimachi, 1981; Moore and Bedford, 1983). This maturation process is referred to as capacitation (Chang, 1951; Austin, 1951). Many of the changes that occur in sperm during capacitation are concerned with plasma membrane alterations, including rearrangements of intramembranous particles, removal of sperm surface components and a decrease in net negative charge.

Invertebrate sperm transferred to females for storage in seminal receptacles have been shown in some cases to undergo maturational changes similar to those occurring in capacitation of mammalian sperm. Such changes have been demonstrated in the shrimp, *Siconia* (Clark *et al.*, 1984). Seminal fluid of the sea urchin, *Arbacia*, reportedly prevents the rapid metabolic decline of sperm. If sperm are added to egg jelly in the presence of seminal fluid respiration and viability are prolonged. It has been speculated that this may be a result of surface modification such that the activity of the spermatozoon is altered (Shapiro and Eddy, 1980). This scheme is similar to capacitation in mammals. Moreover, it indicates that capacitation may not be apparent in some invertebrates due to the rapid dissociation of seminal components from the spermatozoon.

The site of capacitation in mammals appears to vary from one species to another, although successive regions of the female reproductive tract differ in their potential (Yanagimachi, 1981; Moore and Bedford, 1983). In species where semen is deposited in the uterus, capacitation occurs primarily in the oviduct. In species where sperm is deposited in the vagina, it usually starts there and continues in higher regions of the female reproductive tract. The oviduct is more effective than the uterus; however, the fact that capacitation usually occurs most rapidly when sperm are allowed to pass along the entire tract suggests a synergism. Cross-fertilization experiments have demonstrated that the tract of one species is capable of capacitating sperm of a different species.

The sperm of mice, rats, guinea pigs, hamsters, rabbits and humans can also be capacitated in defined media (Yanagimachi, 1981). Since

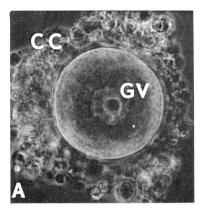
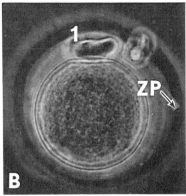

Figure 2.4 A and B (A) Mouse oocyte surrounded by cumulus cells (CC). Because of the density of cumulus cells the zona pellucida is not readily apparent. (B) Mature mouse ovum treated with hyaluronidase to disperse cumulus cells making the zona pellucida (ZP) apparent. GV, germinal vesicle; 1, first polar body.

the mechanism of capacitation is generally believed to be a cell surface phenomenon, the ability of some sperm to capacitate *in vitro* may mean that components of the sperm surface are eluted or are modified in artificial media. Enzymes in the female reproductive tract, blood cells and cumulus cells, have also been implicated as active participants in sperm capacitation in some species.

Since the lifespan of capacitated sperm is limited, it would seem expedient that completion of the process be synchronized temporally and spatially with ovulation. It may have evolved in response to the development of investments (cumulus cells, extracellular matrix and zona pellucida) that the mammalian sperm must penetrate (Fig. 2.4).

Generally the major events of capacitation leading to the acrosome reaction include: an increase in permeability to calcium, modification of the sperm plasma membrane, activation of adenylate cyclase and acrosomal enzymes (Clegg, 1983). Elevated calcium levels may regulate modification of sperm plasma and acrosomal membranes, the activities of acrosin and adenylate cyclase, and sperm motility via calmodulin (Tash and Means, 1982).

Calcium is present in the secretions of the male and female accessory glands and it has been shown that bovine sperm from the cauda epididymus are able to rapidly accumulate exogenous calcium (Babcock, Singh and Lardy, 1979). Sperm calcium uptake is prevented or delayed through the inhibitory action of a protein in seminal fluid that apparently acts upon the sperm plasma membrane. The loss of such a protein may allow capacitating sperm to accumulate calcium. Calcium-ATPase activity has been shown to be associated with membranes of sperm from a number of different species and membrane vesicles derived from sperm actively accumulate calcium (Gordon, Dandekar and Eager, 1978; Bradley and Forrester, 1980). Perhaps this calcium transport system is modulated by seminal plasma proteins.

Calcium involvement in the activation of adenylate cyclase during capacitation has been reviewed (Clegg, 1983). Adenylate cyclase activity increases under conditions that yield capacitated sperm. When guinea pig sperm are added to medium containing calcium they undergo a thirty fold increase in cAMP within 30 seconds; when added to medium lacking calcium only a threefold increase in cAMP is observed (Hyne and Garbers, 1979). D-600, a calcium transport antagonist, blocks the calcium-dependent increase in cAMP. Sperm capacitated in calcium-free medium demonstrate an increase in cAMP 1 min after exposure to calcium-containing media and a maximum number of acrosome reactions within 10 min. Phosphodiesterase inhibitors decrease the time required to obtain the acrosome reaction in calcium-containing media. These results indicate that the mammalian sperm acrosome reaction is associated with both a primary transport of calcium and a calcium-dependent increase in cAMP. Because cAMP analogues do not induce an acrosome reaction in the absence of calcium, the increase in sperm cAMP, induced by additions of calcium, possibly reflects one of a number of calcium-dependent events associated with the acrosome reaction.

Modifications in the sperm plasma membrane as a result of capacitation have been demonstrated by a variety of techniques. Capacitated sperm show a change in lectin binding in the acrosomal region and do not bind specific antibodies derived from uncapacitated sperm surface antigens (Koehler, 1978). Both the changes in antibody and lectin binding may be affected by a masking and/or removal of surface components. Since capacitation can be reversed by exposure to seminal plasma, it is possible that interactions of the seminal plasma and the sperm surface are key features of this process. The removal of 'coating' or decapacitation factors, by either high ionic strength media or by glycosidase digestion, suggests that absorbed components, such as glycoconjugates, on the sperm surface are released coincident with capacitation.

Surface galactosyltransferases on uncapacitated mouse sperm are preferentially loaded with poly-N-acetyllactosamine substrates (Shur and Hall, 1982a). With capacitation in calcium-containing medium

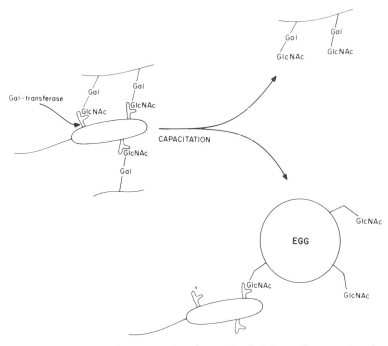

Figure 2.5 Diagram illustrating the release of poly-N-acteyllactosamine glycosides from the sperm surface during capacitation. Glycoconjugate release could be facilitated by either dilution in the oviduct, increased ionic strength, glycosidase digestion or UDP-Gal-mediated catalysis of the galactosyltransferase reaction. Only the terminal disaccharides of the 'decapacitation factors' are illustrated. Reproduced with permission from Shur and Hall (1982a).

these substrates are released from the sperm surface, thereby exposing galactosyltransferase for zona pellucida binding (Fig. 2.5). Galactosylation of endogenous polylactosaminyl substrate is (a) reduced when sperm are incubated in calcium-free medium or treated with antiserum that reacts to galactosyltransferase substrate, and (b) increased by galactosylation of exogenous N-acetylglucosamine and binding to the zona pellucida. When added back to an *in vitro* fertilization, assay glycosides function as 'decapacitation factors', inhibiting sperm–egg binding by competing for sperm surface galactosyltransferase.

Investigations employing freeze-fracture replication have shown that areas of the sperm plasma membrane associated with the acrosome are cleared of intramembranous particles during capacitation (Friend, 1980). The binding of filipin, an agent which indicates the presence of β-hydroxysteroid, also decreases in the plasma membrane of the acrosomal region during this period. Both the changes in intramembranous particle distribution and sterol content of the sperm plasma membrane are believed to be in preparation for the acrosome reaction and render the sperm plasma membrane and the acrosome membrane fusigenic. The mechanisms for these membrane alterations have not been demonstrated. They may represent removal of components along the outer and/or inner surface of the membrane that restrict the mobility of intramembranous particles. Dramatic redistribution of surface antigen, resulting from a migration of molecules originally present on the posterior tail, occurs during capacitation of guinea pig sperm (Myles and Primakoff, 1984). This rearrangement of pre-existing surface molecules may act to regulate sperm functions during fertilization.

Acrosome reaction

Although the necessity of the acrosome reaction as an essential prerequisite for gamete fusion has been questioned by some investigators, extensive studies of fertilization in different species have shown that in forms possessing acrosomes, only acrosome-reacted sperm fuse with

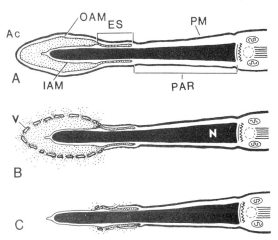

Figure 2.6   Diagram of the acrosomal reaction in the hamster. (A) Before, (B) during the vesiculation of the outer acrosomal (OAM) and plasma membranes (PM) and release of acrosomal contents, and (C) after the acrosomal reaction. Ac, acrosome; PAR, postacrosomal region; N, nucleus; ES, equatorial segment; IAM, inner acrosomal membrane; V, vesicles formed as a result of multiple fusions between the plasmalemma and the outer acrosomal membrane. Lines depicting the plasma and the outer acrosomal membranes have different thicknesses to illustrate their fates subsequent to the acrosome reaction.

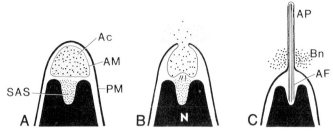

Figure 2.7 Diagram of the acrosome reaction in the sea urchin, *Arbacia*. (A) Intact acrosome (Ac). (B) Exocytosis of acrosomal contents via fusion between the plasma membrane (PM) and the acrosomal membrane (AM). (C) Formation of the acrosomal process (AP) containing polymerized actin (AF) and surrounded by bindin (Bn). SAS, subacrosomal space containing unpolymerized actin; N, nucleus. Lines depicting the plasma and the outer acrosomal membranes have different thicknesses to illustrate their fates subsequent to the acrosome reaction.

the ovum. Among many invertebrates and vertebrates it is generally found that the acrosome undergoes a change when the spermatozoon comes into contact with the ovum (Dan, 1967; Colwin and Colwin, 1967). In mammals, during the acrosome reaction, the plasma membrane overlying the acrosome and the outer acrosomal membrane, fuse at multiple sites and form an array of vesicles (Barros *et al.*, 1967). The vesicles that form have been shown to be mosaics consisting of membrane derived from both the plasma and the outer acrosomal membranes (Russell, Peterson and Freund, 1979). At the level of the equatorial segment, the plasma membrane and the acrosomal membrane fuse to maintain a continuous membrane that delimits the contents of the spermatozoon (Fig. 2.6).

As a result of the acrosomal reaction, the vesicles derived from the fusion of the plasma and acrosomal membranes and the contents of the acrosome are released to the surrounding environment (Fig. 2.6). The release of acrosomal contents in this manner is akin to exocytosis in secretory cells. In addition, the apex and much of the lateral aspect of the sperm head is delimited by membrane derived from the inner acrosomal membrane. Intramembranous particle-free areas are present in the plasma membrane of the equatorial segment and postacrosomal region of guinea pig sperm following the acrosome reaction (Friend, 1980). The development of such areas, which are believed to be relatively rich in lipid, may endow these membranous regions with the capacity to fuse with the egg plasma membrane.

The equatorial segment of mammalian sperm has features that may

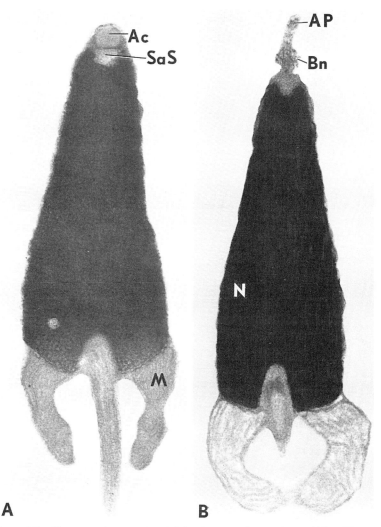

A   **B**

Figure 2.8   Electron micrographs of *Arbacia* sperm that are intact (A) and have undergone the acrosome reaction (B). Ac, acrosome; N, nucleus; M, mitochondria; AP, acrosomal process; Bn, bindin; SaS, subacrosomal space containing unpolymerized actin.

prevent it from participating in the acrosome reaction (Moore and Bedford, 1983). The inner and outer acrosomal membranes in this region of the sperm head have an unusual pentalaminar structure while material located between the membranes is organized in a septate fashion. The plasma membrane associated with the rostral end of the

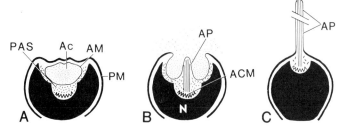

Figure 2.9   Diagram of the acrosome reaction in the starfish, *Asterias*. (A) Intact acrosome (Ac). (B) Exocytosis of acrosomal contents via fusion of the acrosomal membrane (AM) and the sperm plasmalemma (PM) and formation of the acrosomal process (AP) by the polymerization of actin within the periacrosomal space (PAS). The actomere (ACM) nucleates the polymerization of actin. (C) Acrosome reaction completed with the formation of a long acrosomal process containing actin filaments; N, nucleus. Lines depicting the plasma and the outer acrosomal membranes have different thicknesses to illustrate their fates subsequent to the acrosome reaction.

acrosome is rich in anionic lipids. This is consistent with a high membrane fluidity in this area which would be conducive for membrane fusion. In contrast, membrane of the postacrosomal cap, which does not participate in fusion events of the acrosome reaction, has a lower concentration of anionic lipids.

In many of the invertebrates studied, fusion of the plasma and acrosomal membranes is restricted to the apex of the acrosome and consequently, few, if any, vesicles are formed (Dan, 1967; Figs. 2.7–2.9). Concomitantly, in species such as sea urchins, starfish and sea cucumbers, monomeric actin, which in the unreacted spermatozoon is confined to the subacrosomal or periacrosomal region, polymerizes into a core of filaments and an acrosomal process is formed (Tilney, 1975a; Figs. 2.7–2.9). Polymerization of actin filaments is polarized and in the sea cucumber, *Thyone*, the actomere, located at the base of the acrosomal process, acts as a nucleating site for actin polymerization (Tilney, 1978). The length of the acrosomal process that forms can be quite variable: 11 µm in the starfish, *Piaster*, up to 90 µm in *Thyone*, and approximately 1 µm in *Arbacia*. The derivation of the membrane that accommodates the increase in surface area due to the formation of an acrosomal process has not been determined.

In other invertebrate sperm such as the pelecypods, *Mytilus* and *Spisula*, the acrosome surrounds a process containing actin filaments that is then borne with the acrosome reaction (Fig. 2.10). In these cases then, there is no massive polymerization of actin to form an apical

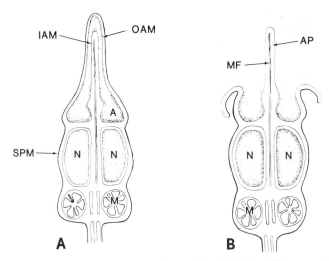

Figure 2.10 Acrosome reaction in the mollusc, *Mytilus*. (A) Sperm with intact acrosome (Ac). (B) Exocytosis of acrosomal contents and exposure of the acrosomal process (AP). Unlike *Arbacia* and *Asterias* sperm in which the acrosomal process is formed via the polymerization of actin, in *Mytilus* actin filaments (MF) are present in the unreacted spermatozoon. IAM and OAM, inner and outer acrosomal membranes; SPM sperm plasma membrane; N, nuclei; M, mitochondria.

projection. In the horseshoe crab, *Limulus*, a long cable of actin filaments is found coiled along the posterior of the sperm nucleus. When the acrosomal vesicle undergoes dehiscence this cable moves through a channel in the nucleus and forms a long acrosomal process that projects from the sperm apex (Tilney, 1975b). The mechanism for the explosive uncoiling of this cable and its projection through the nucleus to the anterior region of the sperm is coordinated with structural rearrangements of its actin components.

The sperm of some organisms, such as teleosts, lack an acrosome. Interestingly, many teleosts possess a small hole or micropyle in the thick covering surrounding the ovum which provides a portal for sperm entry (Fig. 2.11). The presence of a channel through the layer(s) surrounding the teleost egg may be correlated with the absence of an acrosome. However, this relation appears to be more complex as some organisms, such as insects, possess eggs with micropyles and sperm with acrosomes. Micropyles are also found in the eggs of some nemertines, gastropods, pelecypods and cephalopods.

Enzymes reportedly contained within the acrosomes of mammalian sperm, which are externalized with the acrosome reaction and

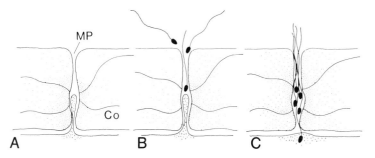

Figure 2.11   Entry of sperm into a sturgeon egg. (A) The micropyle (MP) traverses the chorion (Co) and contains a projection of egg cytoplasm. (B) Sperm traversing the micropyle contact and fuse with the egg. (C) Incorporation of a spermatozoon. Adapted from Austin (1965).

apparently assist the sperm in its movements to, and incorporation into, the egg include: acrosin, hyaluronidase, neuraminidase, cathepsin D, and phospholipase A (Yanagimachi, 1981). Based on the presence of these lytic agents the acrosome may be considered to be a special type of lysosome. Hyaluronidase is localized within the matrix of the acrosome; some is reportedly bound to the acrosomal membrane following the acrosome reaction (Morton, 1975). Hyaluronidase is also detected in the extracellular medium before the acrosome reaction and, hence, is believed to be loosely associated with the spermatozoon. Acrosin is present in the acrosome as a zymogen (proacrosin) and is autocatalytically activated (Meizel and Mukerji, 1975).

   Localization of the trypsin-like enzyme, acrosin, to the acrosome by histochemical procedures include the use of proteinate, fluorescently labeled antiacrosin antibody and horseradish peroxidase-labeled antibody to acrosin (Stambaugh, Smith and Foltas, 1975; Garner *et al.*, 1975; Morton, 1975). The proteolytic activity of mammalian acrosin has been demonstrated by sperm smears on gelatin films (Gaddum-Rosse and Blandau, 1972). Lytic enzyme released from the sperm digests gelatin films containing carbon particles. Consequently, when examined by light microscopy the area immediately surrounding the sperm head is more translucent than other regions and appears as a bright halo. Lytic enzymes are also present in invertebrate sperm (Dan, 1967; Green and Summers, 1980). Trypsin-like enzyme activity is present in sea urchin sperm and released with the acrosome reaction (Levine, Walsh and Fodor, 1978).

   In addition to lytic substances, the acrosomes of sea urchin sperm contain a 30 500 molecular weight protein, called bindin (Vacquier and

Moy, 1977; Vacquier, 1980). A similar acrosomal component has also been identified in sperm from other invertebrates. Bindin from sea urchin sperm has a species preferential affinity for a receptor glycoprotein found on the vitelline layer of unfertilized eggs. It corresponds to material shown in early ultrastructural investigations that coated the acrosomal process and was believed to mediate gamete attachment. Evidence that bindin does, in fact, mediate gamete attachment comes from investigations indicating that antibodies reportedly specific to this protein localize at the site of sperm–egg attachment. Isolated bindin agglutinates unfertilized eggs; the protein is found in the region of contact between eggs. This agglutination is blocked by a glycopeptide fraction from the egg surface suggesting that bindin interacts specifically with molecules (receptors) within the vitelline layer. Whether the acrosomes of mammalian sperm possess a component comparable to bindin has not been determined.

The site of the acrosome reaction in invertebrates is variable and in some cases controversial (Dale and Monroy, 1981). In starfish and the horseshoe crab, the acrosome reaction occurs at the outer margin of a well-defined jelly layer; a long acrosomal process is formed that contacts and makes fusion with the egg plasma membrane (Fig. 2.12). Penetration of the jelly and vitelline layers in these instances is a result of mechanical forces generated by filament extension and lytic products formerly within the acrosome. In sea urchins the acrosome reaction normally takes place when the sperm contacts the vitelline layer (Vacquier, 1979; Aketa and Ohta, 1977).

Timing of the acrosome reaction in the normal course of mammalian fertilization is controversial (Moore and Bedford, 1983). It is generally believed that mammalian sperm complete the acrosome reaction before penetrating the zona pellucida. The most likely sites for this reaction are

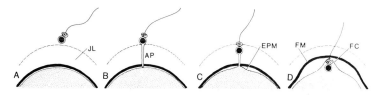

Figure 2.12   Interaction of starfish (*Asterias*) sperm and egg. (A) Approach of the spermatozoon. (B) Acrosome reaction induced by contact with the jelly layer (JL). An acrosomal process (AP) is formed that projects to the surface of the egg. (C) Fusion of the acrosomal process and the egg plasma membrane (EPM). (D) Incorporation of the sperm nucleus within a fertilization cone (FC). FM, fertilization membrane.

within the cumulus and at the surface of the zona pellucida.

The release of lytic agents such as hyaluronidase and trypsin-like enzymes during the acrosome reaction assist the sperm through the cellular and noncellular layers surrounding the mammalian ovum. Intuitively, one might expect that when the sperm approaches, or as it penetrates, these layers it would undergo the acrosome reaction. Although numerous investigations in mammals have been carried out substantiating this point, recent studies indicate that this may not be true for all species. Evidence supporting the idea that mammlian sperm undergo the acrosome reaction as they pass through the cumulus includes the following:

1. The extracellular matrix of the cumulus mass consists of hyaluronic acid which is degraded by hyaluronidase resulting in cumulus dispersal.
2. Acrosomes of mammalian sperm contain hyaluronidase which is released with the acrosome reaction. However, hamster sperm appear to release hyaluronidase prior to the initiation of the acrosome reaction. It is difficult to see how this enzyme can be released from the organelle and the acrosome remain intact.
3. Inhibitors of hyaluronidase activity and antibodies to sperm hyaluronidase block fertilization at the level of the zona pellucida (Dunbar *et al.*, 1976; Perreault, Zaneveld and Rogers, 1979).

Despite these observations, a number of studies indicate that the sperm may penetrate the cumulus intact, bind to the zona pellucida and then undergo the acrosome reaction (Florman and Storey, 1982; Storey *et al.*, 1984). Capacitated hamster sperm with intact acrosomes bind to the zona pellucida whereas sperm that have undergone the acrosome reaction fail to bind. Furthermore, sperm plasma membrane receptors have been identified from the zona pellucida of mouse eggs (Bleil and Wassarman, 1980a,b; 1983). The means by which sperm penetrate the investments of mammalian eggs and the role that acrosomal enzymes may play in this process have not been firmly established.

Egg water, i.e. aqueous solutions in which eggs are incubated to allow their jelly layers to dissolve, and alkaline pH initiate the acrosome reaction of sperm from many marine invertebrates (Dan, 1967). Extracellular calcium is required for this reaction (Schackmann, Eddy and Shapiro, 1978). Calcium uptake in sea urchins occurs via two mechanisms at the time of the acrosome reaction (Schackmann, Eddy and Shapiro, 1978; Schackmann and Shapiro, 1981). The first occurs very early, is sensitive to D-600 (an agent that blocks calcium channels) and

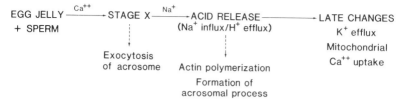

Figure 2.13 Ionic changes associated with the acrosome reaction. See Schackmann and Shapiro (1981).

is associated with a change (X) that allows the acrosome reaction to proceed if other requirements are met (Fig. 2.13). The second is sensitive to respiratory inhibitors and is insensitive to D-600, suggesting that it is mitochondria-derived. The latter follows extension of the acrosomal process and appears to have no role in the acrosomal reaction, although it accounts for the net uptake of calcium that is measured in reacted sperm.

A fucose sulfate component from the jelly layers of sea urchin eggs induces the acrosome reaction (SeGall and Lennarz, 1979; Kopf and Garbers, 1980). There is some species specificity with respect to the ability of this material to stimulate the acrosome reaction. Notably calcium uptake in sperm does not occur when the fucose sulfate component fails to induce an acrosome reaction. Mixing reactive and unreactive jelly layer components does not inhibit the uptake of calcium in affected sperm, suggesting a receptor type of interaction. These observations correlate with those of Summers and Hylander (1975) who showed that heterologous crosses between different sea urchin species induce the acrosomal reaction, although there is no sperm–egg binding or fusion. These results indicate that species specificity of fertilization is not limited to the induction of the acrosome reaction exclusively, but also resides with events that normally follow the acrosome reaction.

The fucose sulfate component of egg jelly induces a calcium dependent activation of adenylate cyclase, and an increase in cAMP (Garbers and Hardman, 1976). The acrosome reaction and increase in cAMP are tightly correlated and temporally consistent with an involvement of cyclic nucleotides in this modification of the acrosome. That the initial step in the acrosome reaction appears to involve an increase in calcium permeability suggests calcium-calmodulin may mediate this reaction (Garbers and Kopf, 1980). Calmodulin is present in human, rabbit, hamster, rooster and sea urchin sperm (Jones *et al.*, 1978). It comprises about 12% of the total soluble protein in rabbit sperm heads (nucleus and acrosome) and is similar to calmodulin from somatic cells in its ability to activate cyclic nucleotide phosphodiesterase.

Sea urchin sperm do not undergo an acrosome reaction at a pH below 7.5 and at pH values greater than 8.8 the reaction is triggered spontaneously in the absence of egg jelly. These observations suggest that there may be a proton efflux during the acrosome reaction which has been confirmed (Schackmann and Shapiro, 1981). When dissolved jelly layer is added to sperm there is a transient release of protons followed by the acrosome reaction (Fig. 2.13). Unlike the acid release which is related to the burst in respiration when sperm are mixed with sea water this is not sensitive to respiratory inhibitors.

Although calcium is required for the exocytosis of the acrosome, it does not appear to be involved in the formation of the acrosomal process. Using the ionophores A-23187 and X-536A Tilney *et al.* (1978) induced actin polymerization and the formation of the acrosomal process in *Thyone* (sea cucumber) and found an acid release in isotonic sodium chloride or potassium chloride at pH 8. At pH 6.5 there is no acid release and no polymerization of actin to form an acrosomal process. The correlation between acid release and actin polymerization suggests that the increase in intracellular pH permits actin to be released from a 'bound' form (i.e., a complex consisting of profilin and non-filamentous actin) and to assemble into filaments.

Sodium and potassium have also been shown to be involved in the acrosome reaction (Schackmann, Eddy and Shapiro, 1978; Schackmann and Shapiro, 1981). Coincident with the acrosome reaction and the proton efflux, sea urchin sperm incubated with egg jelly accumulate sodium. When sodium uptake is prevented, the acrosome reaction and proton release are inhibited, suggesting that fluxes of sodium and hydrogen are linked. The ratio of sodium uptake to proton release is approximately 1:1 suggesting a coordinate change. Inhibitors of potassium flux, as well as increasing potassium concentrations, inhibit the acrosome reaction in sea urchins, indicating that the acrosome reaction is accompanied by an efflux of potassium (Lopo, 1983). In summary, ion fluxes regulating the sea urchin acrosome reaction appear to be linked. Egg jelly induces a calcium uptake in sperm which leads to an exocytosis of the acrosomal vesicle; sodium uptake and acid release follow resulting in an alkalinization of the sperm, the polymerization of actin, and the formation of an acrosomal process (Fig. 2.13).

Two polypeptides, located on the surface of sea urchin (*Strongylocentrotus*) sperm and of 84 000 and 65 000 molecular weight, have been implicated in the regulation of the acrosome reaction (Lopo and

Vacquier, 1980). Antibodies to the 84 000 molecular weight protein block the jelly layer-induced acrosome reaction. The acrosome reaction can be induced when antibody treated sperm are incubated with the calcium ionophore A-23187 at pH 9.2 indicating that the inhibition is due to the blockage of a molecule playing a role in the acrosome reaction. Univalent (Fab) antibodies to an 85 000 protein from sperm of the sea urchin *Arbacia*, inhibit the acrosome reaction and fertilization (Saling, Eckberg and Metz, 1982). A-23187 or the addition of a 68 000 molecular weight polypeptide derived from sperm membrane preparation eliminates the inhibitory activity of the Fab fragments. Results of these studies suggest that both the lower molecular weight (64 000/68 000) and the higher molecular weight polypeptides (84 000/85 000) play a different but necessary role in the acrosome reaction of sea urchins.

Studies of sperm–egg interactions in mice have demonstrated that zona pellucida glycoprotein, ZP-3, serves as both a receptor for sperm and an inducer of the acrosome reaction (Bleil and Wassarman, 1983). Proteolysis of ZP-3 results in its inability to induce the acrosome reaction, suggesting that the polypeptide chain of ZP-3 plays a role in this process. Follicular fluid and cumulus cells have also been shown to stimulate sperm to undergo the acrosome reaction in hamsters. However, in some mammals fertilization of washed ova will occur *in vitro* or *in vivo* in the absence of follicular products suggesting that the 'factor' necessary to trigger the acrosome reaction is not derived from the egg. Consistent with these observations are studies that demonstrate the presence of motile, oviductal, acrosome-reacted sperm in the absence of ovulation. Furthermore, numerous *in vitro* systems have been shown to affect the acrosome reaction and include parameters such as pH, temperature, energy source, albumin, taurine, and epinephrine, as well as others (Yanagimachi, 1981). In general most *in vitro* systems involve the removal of a substance (decapacitation factor) from the mammalian sperm surface, and initiate an influx of calcium which affects membrane fusion and the conversion of proacrosin to acrosin.

3

Sperm–egg binding

Numerous observations of sea urchins have demonstrated the presence of sperm receptors in the egg vitelline layer, a carbohydrate-rich extension of the plasma membrane that is homologous with the glycocalyx of somatic cells. Sea urchin eggs treated with proteolytic enzymes, such as trypsin, show a decrease in fertility due to a reduction in sperm–egg binding (Schmell *et al.*, 1977). Unfertilized eggs incubated in cortical granule protease also undergo a decrease in the ability of sperm to bind, resulting from an alteration in receptors (Vacquier, Tegner and Epel, 1973; Carroll and Epel, 1975). Large glycoproteins isolated from the vitelline layers of sea urchin eggs show a species specific inhibition of fertilization (Aketa, 1973; Schmell *et al.*, 1977; Glabe and Vacquier, 1978; Glabe and Lennarz, 1981). Protein isolated from the surface of sea urchin (*Hemicentrotus*) eggs binds homologous sperm and antibody directed to it blocks fertilization (Aketa and Tsuzuki, 1968; Aketa and Onitake, 1969). Membrane protein preparations from one species of sea urchin (*Arbacia*) eggs inhibit fertilization by sperm of the same species but have no effect on the ability of sperm from a different species (*Strongylocentrotus*) to fertilize homologous eggs.

In the presence of soluble components derived from the egg, acrosome reacted sperm exhibit a species specific binding to each other, suggesting that specific receptors are present on the ovum surface that facilitate gamete binding (Kinsey, Rubin and Lennarz, 1980). Sperm bind only to the external surface of isolated vitelline layers and also exhibit saturation kinetics consistent with the notion of a sperm receptor on the ovum surface (Vacquier and Payne, 1973). Unfertilized sea urchin eggs treated with trypsin or cortical granule proteases are not agglutinated by bindin, indicating that a specific sperm receptor on the egg surface is destroyed by proteolysis (Vacquier and Moy, 1978).

Penetration of the vitelline layer is a requirement in order for sea urchin sperm to fuse with the egg plasma membrane. This may occur mechanically by the action of the sperm flagellum with the acrosomal process acting as a spear. In addition, portions of the vitelline layer may be digested by serine proteases present in the acrosome and externalized by the acrosome reaction. Trypsin-like activity has been localized to the acrosomal process of sea urchin sperm and is believed to be involved in vitelline layer penetration (Green and Summers, 1980). Fertilization decreases in the presence of the protease inhibitor, tosyl-phenylalanine-chloromethyl-ketone (TPCK). If the vitelline layer is removed, TPCK is not inhibitory suggesting that normally the vitelline layer is digested by acrosome-derived enzymes.

Further evidence that hydrolases participate in the penetration of the vitelline layer comes from investigations with other invertebrates. Ascidian ova are enveloped by a complex of follicle cells, chorion and test cells that presumably presents a formidable barrier to successful gamete fusion. The entire surface of the *Ciona* chorion is apparently available for sperm binding and, despite lacking a typical acrosome, ascidian sperm are able to penetrate these barriers (De Santis, Jammuno and Rosati, 1980). Trypsin and chymotrypsin inhibitors block the fertilization of ascidian eggs (Hoshi, Numakunai and Sawada, 1981). However, if the chorion is removed, the egg can be fertilized in the presence of inhibitors.

Investigations of sperm–egg binding in mammals have concentrated, for the most part, on the interaction of the sperm and the zona pellucida. The interaction of the mammalian sperm with the zona pellucida may take several forms. That which is easily disrupted by pipetting, occurs at 2 or 37°C, is not species specific, and is generally referred to as 'attachment' (Hartmann, Gwatkin and Hutchison, 1972). After a short period a tenacious union occurs, 'binding', which is species specific, and calcium- and temperature-dependent (Gwatkin, 1977). Binding in hamsters is not affected by neuraminidase, lysozyme, α- and β-amylase, glucoamylase, or β-glucuronidase, but it is trypsin sensitive.

Mouse sperm possess galactosyltransferase which is activated during capacitation (Shur and Hall, 1982a,b). Galactose residues are lost from the active site of the enzyme so allowing it to bind to the zona pellucida. This binding may occur at the N-galactose residues of zona pellucida-glycoprotein (ZP-3) which Bleil and Wassarman (1983) proposed is the zona recognition molecule for sperm.

In vitro studies with cumulus-intact and cumulus-free mouse eggs demonstrated the presence of receptors on the sperm plasma membrane for zona components (Bleil and Wassarman, 1983; Storey *et al.*, 1984). Acrosome intact sperm bind to the zona pellucida at glycoprotein,

ZP-3. Two other glycoprotein components of the zona pellucida (ZP-1 and ZP-2) have no apparent effect on sperm binding. ZP-3 from fertilized eggs does not have sperm receptor activity which is consistent with the inability of dissolved embryo zonae pellucidae to inhibit binding of sperm to eggs *in vitro* and the inability of sperm to bind to fertilized eggs. There is no change in the electrophoretic mobility of ZP-3 from fertilized eggs suggesting that a subtle change, possibly one involving the carbohydrate portion of the glycoprotein, occurs upon fertilization.

Although the binding of sperm to the zona pellucida and induction of the acrosome reaction in mice appear to be mediated by ZP-3 alone, these two events can be distinguished from one another (Florman, Bechtol and Wassarman, 1984). The sperm receptor activity of ZP-3 is dependent upon its carbohydrate content, whereas acrosome reaction-induced activity is dependent upon the polypeptide chain of ZP-3 as well. In this connection it has been reported that exposure of zonae pellucidae to exudate from activated eggs, presumably cortical granule contents, results in a decrease in sperm binding. Since this effect is prevented by trypsin inhibitors, it has been suggested that a cortical granule-derived protease inactivates sperm receptors in the zona pellucida. While there is no evidence for proteolysis of ZP-3 following fertilization, it is possible that a small glycopeptide or a specific glycosidase is released during the cortical granule reaction which inactivates sperm receptors.

These observations, together with studies demonstrating an inhibition of binding when eggs are treated with antizona antibody, some lectins and trypsin, are consistent with the view that a sperm receptor is present in the zona pellucida (Gwatkin, 1977). Additionally, investigations with inhibitors of glycoprotein synthesis, lectins and monosaccharides indicate that sperm surface carbohydrates may also be required for different stages of fertilization, including sperm–egg fusion (Ahuja, 1982).

Penetration of the zona pellucida requires approximately 4 min in hamsters and about 20 min in mice (Yanagimachi, 1981). The movement of the sperm through the zona pellucida appears to be facilitated by a bobbing movement of the head. Lytic agents from the acrosome are also believed to participate in penetration because trypsin inhibitors block fertilization. Despite this evidence there are doubts as to the specificity of this inhibition as relatively high concentrations of inhibitors are required to block fertilization. Furthermore, it is uncertain whether the acrosome reaction can take place in the presence of the inhibitors. Hence, failure of inhibitor-treated sperm to penetrate the zona pellucida and fertilize may be due to a blockage of the acrosome reaction.

4

Gamete fusion and sperm incorporation

The site of gamete fusion and sperm incorporation in many eggs is restricted to a specific part of the ovum. Eggs with micropyles, such as fish and insects, are highly specialized in this respect as a portal in the layers investing the egg permits access of the sperm to a specific region of the ovum. The egg plasma membrane beneath the micropyle is the region where the sperm normally fuses with the egg, although experiments with dechorionated eggs of some species show that sperm can fuse at any site along the egg surface. In a number of animal eggs (e.g., cnidarians, hydrozoans and siphonophores) the sperm fuses with the egg near the site of polar body formation (Freeman and Miller, 1982). In the amphibian, *Discoglossus*, sperm enter at a depressed region on the animal pole called the animal dimple (Campanella, 1975). This restriction may be due to the production of a substance that attracts or traps sperm at the site of fertilization, or the presence of a differentiated surface that limits sperm—egg fusion to one site along the ovum surface. The fact that sperm appear to enter both at the animal and vegetal poles in some urodele eggs may be related to the polyspermy that occurs in this species. In the frog, *Rana pipiens*, sperm penetrate the jelly and vitelline layers and apparently approach the plasma membrane anywhere along the perimeter of the egg. Sperm entry, however, occurs only in the animal half of the egg (Elinson, 1980). Sperm incorporation in mammals, particularly the mouse and hamster, occurs along the region containing microvilli, away from the microvillus-free area in which the meiotic apparatus is located (Nicosia, Wolf and Inoue, 1977; Yanagimachi, 1981; Fig. 4.1). In sea urchins, fertilization apparently occurs anywhere along the surface of the ovum (Schroeder, 1980).

Rabbit sperm after passing through the zona pellucida appear to move vigorously in the perivitelline space before attaching to the plasma membrane; 20 minutes later the entire sperm is located within

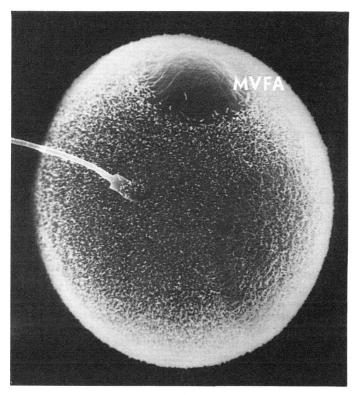

Figure 4.1 Scanning electron micrograph of a hamster spermatozoon fusing with an egg. Zonae-free hamster eggs were inseminated *in vitro* with acrosome-reacted sperm and fixed about 10 min later. The site of gamete fusion is restricted to the area of the egg projected into microvilli. The meiotic spindle is located in the elevated microvillus-free area (MVFA). See Yanagimachi (1981).

the ovum. Hamster sperm attach to the egg plasma membrane apparently without 'wandering' within the perivitelline space. In the meantime the beating sperm tail vibrates and rotates the egg within the perivitelline space. Finally, when the entire sperm is in the perivitelline space its vigorous movements cease.

Participation of microvilli in gamete fusion has been observed with transmission and scanning electron microscopy in hamsters (Yanagimachi and Noda, 1970a; Shalgi and Phillips, 1980a,b). Association of the acrosomal process and its fusion with the egg surface is believed to be mediated at microvilli in sea urchins (Vacquier, 1980; Figs. 4.2 and 4.3). Gamete membrane fusion in the molluscs, *Mytilus* and *Spisula*, is believed to occur in the same manner. In the annelid, *Chaetopterus*, the

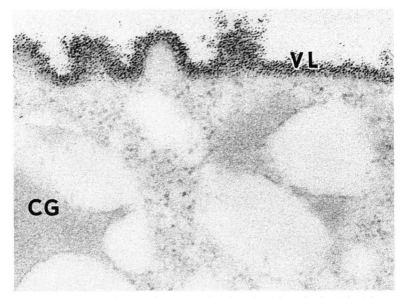

Figure 4.2 Sea urchin (*Arbacia*) egg incubated with cationized ferritin to demonstrate the vitelline layer (VL). CG, cortical granule.

morphology of the egg and the manner in which the gametes interact demonstrate rather conclusively that, at least in this species, sperm–egg fusion is mediated at the tip of the egg's microvilli. Initial contact of sperm and egg involves an unreacted acrosome and the apices of microvilli which penetrate the vitelline layer. Apparently contact of the acrosome with the microvilli initiates the acrosome reaction; subsequently, the egg plasma membrane delimiting the apex of a microvillus fuses with the former inner acrosomal membrane (Anderson and Eckberg, 1983).

Relatively little is known about processes accompanying actual fusion of the gamete membranes. It represents a merging of two membrane domains that results in a confluence of the protoplasmic contents of both the egg and sperm. Changes are believed to occur at the acrosomal reaction that render the sperm fusigenic with the ovum (Colwin and Colwin, 1967) and fusion is likely to be similar at the molecular level to that postulated for other cells. At the site of fusion, the development of membranous changes that accompany the acrosome reaction in guinea pig sperm may occur where intramembranous particle-free areas that are believed to be fusigenic emerge within the plasma membrane (Friend, 1980).

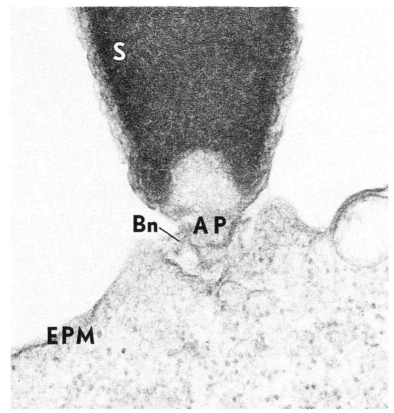

Figure 4.3   Acrosome reacted sea urchin (*Arbacia*) sperm in contact with the egg surface. AP, acrosomal process; Bn, bindin; EPM, egg plasma membrane; S, sperm nucleus.

The precise moment of gamete fusion, i.e. the time at which sperm and egg plasmalemmae become continuous and the cytoplasms of both gametes are able to merge is important in defining possible mechanisms and the chronology of early events of fertilization and their possible causal relation. Correlative ultrastructural and electrophysiological studies of sperm–egg interactions in the sea urchin *Lytechinus* indicate that egg activation, i.e. the initiation of the activation or fertilization potential, precedes while exocytosis of cortical granules commences following sperm and egg plasma membrane fusion.

Fusion of mammalian sperm and egg is in striking contrast to gamete fusion in the marine invertebrates (Colwin and Colwin, 1967). In many of the invertebrates studied so far, gamete fusion occurs between the egg plasma membrane and the tip of the acrosomal process, which is delimited by membrane derived from the acrosomal vesicle (Fig. 4.3). In mammals segments of plasma membrane over the equatorial segment (Bedford and Cooper, 1978) or the postacrosomal region are the first to fuse with the egg plasma membrane (Yanagimachi and Noda, 1970b; Fig. 4.4).

The presence of calcium is reportedly essential for sperm–egg fusion in mammals (Yanagimachi, 1981). In sea urchins calcium dependency for gamete fusion is controversial. According to Sano and Kanatani (1980) activated sperm added to eggs in the presence of the calcium chelator, EGTA are unable to fertilize. However, other investigators have shown that sea urchin eggs can be fertilized in 'calcium-free media' ($<10^{-8}$M; Chambers, 1980; Schmidt, Patton and Epel, 1982). The fusion of sperm and eggs in the absence of calcium is surprising in light of experiments showing that this ion is involved in membrane fusion processes of other cells. Sperm–egg fusion may be an exception; alternatively calcium may be involved but not essential.

That all of the sperm plasma membrane is incorporated into the egg plasma membrane at fertilization is more or less assumed in many instances, although experimental evidence has not verified this unequivocally. Electron microscopic studies of sperm incorporation in some invertebrates and mammals have demonstrated membranous elements, such as vesicles, at the site of gamete fusion that appear to be derived from the fused sperm and/or egg plasmalemmae (Colwin and Colwin, 1967; Bedford and Cooper, 1978).

Investigations examining the integration of the sperm and egg plasma membranes at fertilization, where one of the gametes has been labeled, have been carried out in both invertebrates and mammals (Yanagimachi *et al.*, 1973; Gabel, Eddy and Shapiro, 1979; Longo, 1982). Prior to sperm–egg fusion in hamsters the sperm plasma membrane of the postacrosomal region does not bind colloidal iron hydroxide. Once gamete fusion has been initiated, however, the former sperm plasma membrane is able to bind this marker. The rapid increase in colloidal iron hydroxide binding to the incorporating sperm head is believed to be a result of intermixing of sperm–egg membrane components comparable to the intermingling of antigenic determinants after fusion of somatic cells. Such results are not unexpected for intrinsic glycoprotein and glycolypids intercalated in a fluid bilayer, are freely diffusible in the plane of the membrane. These observations, however, do not exclude the possibility that colloidal iron hydroxide binding receptors are enzymatically added to sperm plasma membrane oligosaccharides after

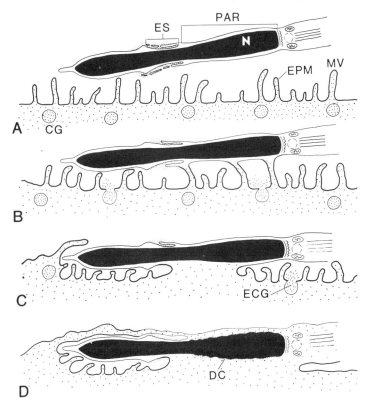

Figure 4.4   Diagramatic representation of a hamster sperm fusing with the egg plasma membrane (EPM). (A) Interaction of the sperm with the egg microvilli (MV). (B) Fusion of microvilli with the postacrosomal region (PAR) of the sperm plasma membrane. (C) and (D) Subsequent stages of sperm incorporation. ES, equatorial segment; CG; cortical granules; ECG, exocytosing cortical granules; N, sperm nucleus; DC, dispersing chromatin. Lines depicting the plasma membranes of the sperm and egg and cortical granules have different thicknesses to illustrate their fates subsequent to gamete fusion. See Yanagimachi and Noda (1970b).

fusion or that colloidal iron hydroxide-binding membrane components are inserted into the sperm plasma membrane following fertilization.

Similar experiments have been carried out with the surf clam, *Spisula* in which concanavalin A binding to the egg, but not the sperm plasma membrane, has been demonstrated by the horseradish peroxidase-diaminobenzidine reaction (HRP-DAB; Longo, 1982). Because of this dichotomy in lectin binding, changes in the affinity of the sperm plasmalemma following its fusion and integration with components of

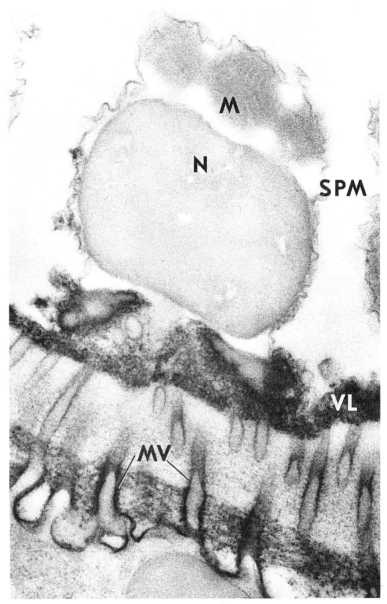

Figure 4.5 Surf clam (*Spisula*) sperm attached to the surface of an egg. The gametes were treated with the lectin, concanavalin A, followed by horseradish peroxidase and then reacted with H_2O_2 and diaminobenzidine. Dense reaction material is present along the surface of the egg microvilli (MV) and the vitelline layer (VL). The sperm plasma membrane (SPM) lacks reaction product due to the absence of concanavalin A binding. Because of this dichotomy in concanavalin A binding integration of the sperm and egg plasma membranes subsequent to gamete fusion may be followed. N, sperm nucleus; M, mitochondria. See Longo (1982).

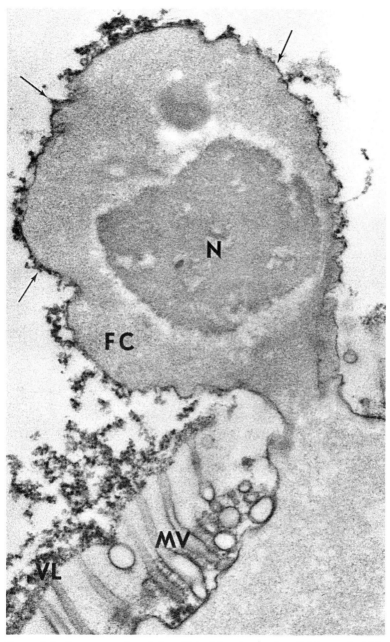

Figure 4.6 Incorporated surf clam (*Spisula*) sperm nucleus (N) located within a fertilization cone (FC). The membrane at the arrows is derived from the spermatozoon and possesses concanavalin A receptors due possibly to their lateral movement within the plane of the egg plasma membrane into the sperm plasma membrane. MV, microvilli; VL, vitelline layer. See Longo (1982).

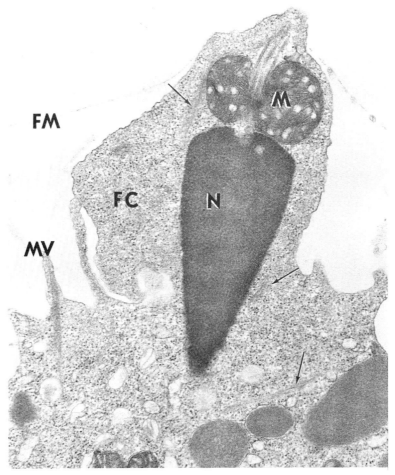

Figure 4.7   Incorporated sea urchin (*Arbacia*) sperm within a fertilization cone (FC). The filamentous structures shown at the arrows contain actin. FM, fertilization membrane; M, sperm mitochondrion; N, sperm nucleus; MV, microvillus containing microfilaments. See Longo (1980).

the egg plasma membrane can be followed (Fig. 4.5). The plasma membranes of fertilized *Spisula* eggs react with concanavalin A-HRP-DAB and are associated uniformly with enzymatic precipitate – except at sites of sperm incorporation by 1 min postinsemination. These portions of unstained plasma membrane are derived from the sperm and associated with the apex of the fertilization cone. From 2 or 4 min postinsemination HRP-DAB reaction product gradually becomes

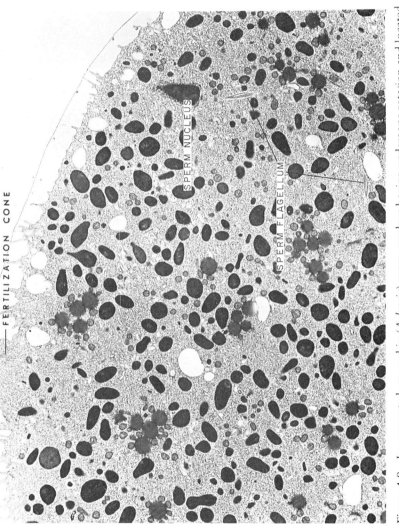

Figure 4.8 Incorporated sea urchin (*Arbacia*) sperm nucleus having undergone rotation and located lateral to its site of entry, the fertilization cone. See Longo (1973).

associated with all the membrane delimiting the fertilization cone (Fig. 4.6). By 4 min postinsemination no difference in staining of plasma membranes derived from the egg or the sperm is detected. These observations are consistent with the movement of concanavalin A binding sites from the egg plasmalemma into the sperm plasma membrane.

Not all components of the sperm and egg plasma membrane appear to intermix rapidly following gamete fusion. Sea urchin and mouse eggs fertilized with fluorescent- or ^{125}I-labeled sperm retain a topographically mosaic surface, as if the lateral mobility of sperm plasma membrane components were restricted and retained as a discreet patch (Gabel, Eddy and Shapiro, 1979). Similar experiments have indicated that labeled sperm plasma membrane components are also internalized after fertilization (Gundersen, Gabel and Shapiro, 1982).

At the site of gamete fusion in marine invertebrate eggs a protuberance forms which has been referred to as the fertilization or incorporation cone (Fig. 4.7). Formation of this structure consists of a movement of egg cytoplasm into the region surrounding the sperm nucleus, mitochondria, and axonemal complex, resulting in a protrusion at the site of sperm entry (Longo, 1973). As the fertilization cone increases in size, the incorporated sperm components move through it into the ovum's cortex. In many organisms the migration of the sperm nucleus into the egg cortex is accompanied by its rotation of 180° (Fig. 4.8). As a result of this turing movement, the sperm nucleus usually becomes located lateral to the site of its entry with its apex directed to the egg plasma membrane.

The maximum size of fertilization cones varies depending upon the organisms in question; in mature sea urchin (*Arbacia*) eggs they measure approximately 6 μm in length by 4 μm in diameter, 5 to 7 min postinsemination. They then regress and are reabsorbed by 10 min postinsemination. Interestingly, in *Arbacia* the fertilization cones that form on immature eggs are much larger than those that develop on mature ova, e.g., sizes of 25 μm in length by 10 μm in diameter are not unusual.

In sea urchin eggs fertilization cones are filled with numerous bundles of actin filaments that show a polarity when reacted with heavy meromyosin or the S1 fragment of myosin (Tilney and Jaffe, 1980). With S-1 decoration the arrowhead complexes that form on the actin filaments in the fertilization cone are directed towards the center of the egg. The microfilaments found in the fertilization cone are believed to polymerize *in situ* from cortical, monomeric actin. Whether and how actin in the fertilization cone might function to effect sperm incorpor-

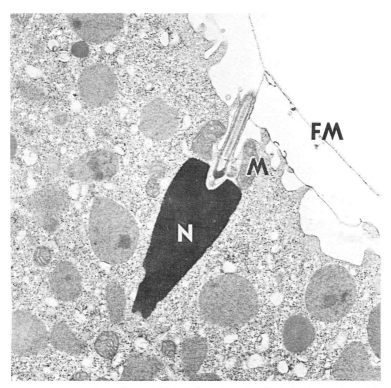

Figure 4.9 Inseminated sea urchin (*Arbacia*) egg in which the formation of the fertilization cone has been inhibited with cytochalasin B. M, sperm mitochondrion; N, sperm nucleus; FM, fertilization membrane. See Longo (1980).

ation is not entirely clear. The actin filaments of the sperm acrosomal processes are also polarized with the heavy meromyosin or S1-actin arrowheads pointing to the sperm nucleus. Consequently, egg myosin could not bridge sliding actin filaments of both the fertilization cone and the acrosomal process to bring about sperm nucleus incorporation; both sets of actin filaments are polarized in the wrong direction when compared to the orientation of myosin and actin of a sarcomere. It is possible that actin filaments present in fertilization cone might be involved in its elevation and enlargement.

Cytochalasin B, a drug that disrupts actin microfilaments, has been shown to inhibit surface activity of fertilized sea urchin eggs, such as microvillar elongation and fertilization cone formation (Longo, 1980; Fig. 4.9). Cytochalasin B treated eggs undergo a cortical granule reaction, elevate a fertilization membrane, and are metabolically activated.

Treatment of eggs with cytochalasin B after sperm incorporation does not significantly affect migration or formation of the male pronucleus. These observations are consistent with the suggestion that at the site of gamete fusion there is a localized polymerization of actin that participates in the formation of the fertilization cone. In addition, experiments with cytochalasin B also indicate that treated sea urchin eggs can be activated by sperm but sperm fail to enter the egg (Gould-Somero, Holland and Paul, 1977; Longo, 1978a). How the sperm is capable of activating the egg in this instance without entering it has not been determined. It is possible that the acrosomal process fuses with the egg plasma membrane, but since actin and its polymerization are impaired, the 'bridge' linking the fused sperm and egg is weak and the sperm is removed from the egg surface by exocytosing cortical granules. Another possibility is that cytochalasin B inhibits fusion of the egg and sperm and that gamete contact/binding in this instance is sufficient for egg activation (Longo *et al.*, 1986).

At the site of sperm entry in anurans a microvillus-free bleb of cytoplasm forms, presumably functionally equivalent to a fertilization cone (Picheral, 1977). Eventually it disappears and is replaced by a small clump of elongate microvilli. The microvillus-free bleb may be pinched off leaving the microvilli. If this is the case then it is possible that plasma membrane components, as well as other sperm-derived structures may be eliminated from the egg. The site of sperm entry reportedly remains detectable as a clump of microvilli for at least 2 h.

In mammals (rats, rabbits and hamsters) following the fusion of the egg and sperm plasma membranes, tongues of cytoplasm surround the anterior portion of the sperm head so forming a vesicle that is present for a time within the zygote (Yanagimachi and Noda, 1970a). At the site of fusion in mammals a protrusion of cytoplasm forms which is homologous to fertilization cones seen in invertebrate eggs and is often referred to as an incorporation cone (Zamboni, 1971). In mouse eggs the protrusion is filled with cytoplasmic organelles found in other regions of the zygote, and along its plasmalemma there is a prominent layer of actin (Maro *et al.*, 1984). In this protrusion the early events of male pronuclear development take place.

Meiotic stages of eggs at insemination

Wilson (1925) indicated that factors which govern the association of the male and female pronuclei may be closely but not necessarily correlated with the relation between the meiotic stage of the egg and the

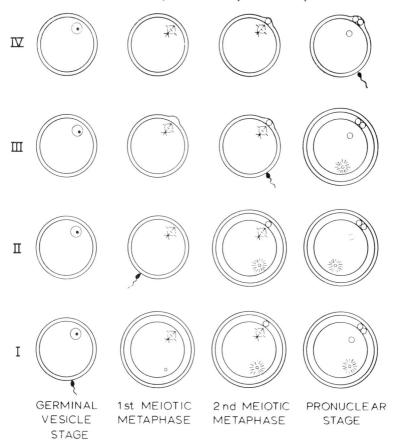

Figure 4.10   Classification of eggs according to the meiotic stage at which they are inseminated.

time it is normally inseminated. Essentially four types of eggs are recognized with respect to their state of meiosis and the time of their insemination (Fig. 4.10). Generally fertilization follows ovulation and eggs are inseminated during or following their meiotic divisions. Several exceptions to this rule exist and include various species of marine annelids and free-living flatworms in which the spermatozoon enters the egg prior to ovulation. Aside from these exceptions the classification that has been formulated is not particularly rigid with respect to animals comprising each class and the classes themselves. For example, in some cases, eggs are inseminated at the first or second anaphase of meiosis. Eggs of some vertebrates are inseminated at meiotic prophase, i.e. the germinal vesicle stage. In other animals, e.g. starfish, a number of classes may apply (II, III or IV). Eggs which comprise class I are fertilized at meiotic prophase and are found in various nematodes, molluscs, annelids and crustaceans. The eggs of some molluscs, annelids and insects are inseminated at the first metaphase of meiosis and make up class II, while the eggs of most vertebrates are found in class III and fertilize at the second metaphase of meiosis. Eggs belonging to class IV are those of various echinoids and coelenterates and are inseminated at the completion of meiosis – the pronuclear stage.

In an examination of this classification two features may be noted. First, there is no apparent phylogenetic relationship between the different groups and the animals comprising them. Second, eggs belonging to classes I, II and III have not completed meiosis and are in a state of meiotic arrest at the time of insemination. Therefore, in these forms, the interaction of the sperm and egg triggers processes that initiate the resumption of meiosis. In these instances fertilization also includes various stages of meiotic maturation.

Changes in the ability to fertilize during the course of meiotic maturation have been investigated in both vertebrates and invertebrates. Generally such studies indicate that sperm can penetrate oocytes at early stages of maturation and that male pronuclear development is usually retarded. The incidence of polyspermy is often high in these cases, indicating that mechanisms underlying the development of the block to polyspermy occur during oocyte maturation. For example, in sea urchins a close association of the cortical granules with the plasma membrane is important for their exocytosis (the cortical granule reaction) and the prevention of polyspermy. That immature sea urchin eggs do not undergo a cortical granule reaction and are polyspermic at low sperm concentrations is consistent with observations demonstrating that their cortical granules do not move to the plasma membrane until meiotic maturation is completed.

5

Egg activation: Ion changes

The resting potential of sea urchin eggs as measured with microelectrodes falls into two general ranges: (1) -5 to -20 mV, which may be a consequence of current leakage, and (2) -60 to -80 mV (Shen, 1983). The resting potential has also been determined by tracer flux experiments and shown to be about -70 mV. Two resting potentials have also been reported for starfish, the lower of which may also be due to current leakage. The resting potentials of a number of invertebrate eggs is given in Table 5.1.

Table 5.1   Resting potentials for eggs of some marine invertebrates in sea water. Data compiled from Shen (1983).

Organisms	Resting potential (mV)
Renella (coelenterate)	-70
Urechis (echiuroid)	-33
Spisula (mollusc)	-20
Ilyanassa (mollusc)	-1 to -20
Dentalium (mollusc)	-70
Strongylocentrotus, Lytechinus (echinoid)	-60 to -80
Halocynthia, Ciona (tunicate)	0 to -20

At insemination in marine invertebrate eggs there is a transient depolarization which has been extensively analyzed in echinoids (Whitaker and Steinhardt, 1982; Shen, 1983; Fig. 5.1). This represents the earliest response of the egg to sperm. There is a positive change in membrane potential in eggs of the sea urchin, *Strongylocentrotus* that goes to $+20$ mV and persists for about one minute before repolarization (Jaffe, 1976). The initial event is an activation of voltage-dependent calcium channels which depolarize the membrane. This transient depolarization is referred to as the fertilization potential. Both the activation of voltage-dependent calcium channels and the duration

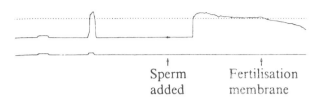

Figure 5.1 Activation potential of a monospermic sea urchin (*Strongylocentrotus*) egg. The top trace is voltage against time; the bottom trace is current against time. The dotted line indicates 0 mV. Reproduced with permission from Jaffe (1976).

of depolarization result from a slow transient increase in sodium permeability of the egg plasma membrane.

The depolarization of the egg is related to the initiation of the rapid block to polyspermy postulated by earlier investigators of fertilization. Support for this contention comes from the following observations (Jaffe, 1976).

1. Eggs with a fertilization potential greater than 0 mV are *not* polyspermic; ova with a more negative fertilization potential *are* polyspermic.
2. If current is applied to eggs to hold the membrane potential to +5 mV, fertilization is blocked; when the current is released, the membrane potential returns to normal and the eggs fertilize.
3. If current is applied to eggs so that the fertilization potential remains below −30 mV, the eggs become polyspermic.

These observations suggest that an initial, rapid block to polyspermy coincides with an electrical depolarization of the membrane.

The ionic mechanisms underlying the fertilization potential exist in the egg plasma membrane since similar potential shifts are elicited by parthenogenetic agents (Steinhardt and Epel, 1974; Chambers, Pressman and Rose, 1974). Correlative ultrastructural and electrophysiological studies (Longo *et al.*, 1986) suggest that a process of gamete interaction, following sperm attachment and prior to sperm-egg fusion, initiates the opening of ion channels.

There is a 1.7-fold change in membrane capacitance at fertilization in sea urchin eggs which may be a reflection of a change in membrane surface area due to the cortical granule reaction (Jaffe, Hagiwara and Kado, 1978). The latent period of 30–45s between the rise of the fertilization potential and the increase in membrane capacitance correlates with the period taken by the cortical granule reaction. This latency also indicates that the rising phase of the fertilization potential is not a consequence of the cortical granule reaction.

Voltage clamping of sea urchin eggs at -70 mV does not prevent development nor does activation occur when unfertilized sea urchin eggs are depolarized. On the basis of these observations it has been suggested that in sea urchins the potential changes at fertilization may not be important for the initiation of development (Shen, 1983). However, ionic currents associated with the fertilization potential and changes in intracellular ion activities are important for triggering development in sea urchins and also in eggs of other marine invertebrates (Chambers, 1976). Potassium, magnesium, and calcium are reportedly not required for development, although divalent cations may be necessary for sperm–egg fusion (Schmidt, Patton and Epel, 1982). The sodium flux at fertilization is crucial since its removal immediately after insemination does not permit development (Chambers, 1976).

In amphibian eggs the potential changes caused by fertilization are due to a chloride flux. The maximum value is reached within 1–2 s and then decreases over a 20–30 min period. As in the case of sea urchins, if the membrane potential is increased, fertilization can be prevented (Cross and Elinson, 1980); eggs can be made polyspermic if the hyperpolarization that normally accompanies fertilization is prevented. The kinetics and duration of membrane potential changes in anurans have been related to the kinetics and duration of the cortical granule reaction. The membrane potential changes are rapid (1–2 s in duration) but transitory. The cortical granule reaction is much slower (from 39–117 s postinsemination) but permanent (Schmell, Gulyas and Hedrick, 1983). Both the membrane potential change and the cortical granule reaction at fertilization in anurans constitute blocks to polyspermy.

Direct experimental evidence for a block to polyspermy at the level of the plasma membrane in mammals is limited. The observation that many sperm can penetrate the zona pellucida of a rabbit egg, yet the ovum remains monospermic, is generally believed to indicate the presence of a block to polyspermy at the level of the plasma membrane. Membrane potential changes are found in hamster eggs which consist of recurring hyperpolarizations beginning with the approximate time of sperm entry. The relation of the hyperpolarization to ionic changes and the block to polyspermy has not been determined. A slow transient depolarization is recorded for fertilized rabbit eggs (McCulloh, Rexroad and Levitan, 1983). The small amplitude of responses compared with the large variations of the resting potentials suggests that the depolarization may be sufficient to block polyspermy. Except for oscillations, the membrane potential of fertilized mouse eggs is constant up to 60 min postinsemination, suggesting that the polyspermy block that is established during this period is not electrically mediated (Jaffe, Sharp and Wolf, 1983). Polyspermy prevention in the mouse, and possibly in other mammals, occurs at several levels including restrictions in the number of sperm reaching the site of fertilization, blocks at the zona pellucida and at the plasma membrane, and elimination of supernumerary sperm (Yu and Wolf, 1981).

The molecular mechanism by which changes in membrane potential alter sperm attachment and fusion has not been determined. It is presumed that egg plasma membrane topography is altered and prevents further sperm interaction. In surf clam (*Spisula*) eggs there is no cortical granule reaction or elevation of a fertilization membrane, yet a complete block to polyspermy is established within 15 s of insemination. It has been suggested that the change in intramembranous particle density of *Spisula* egg plasma membranes at insemination may be related to the development of a block to sperm incorporation (Longo, 1976a).

Mazia (1937) showed that in sea urchin (*Arbacia*) eggs the free calcium content increased upon fertilization. Total calcium content was unchanged, indicating that the concentration of free calcium increased at the expense of the bound form. Using calcium-45, fluxes of calcium following fertilization in sea urchins have been demonstrated (Azarnia and Chambers, 1976; Paul and Johnston, 1978). Injection of micromolar calcium induces egg activation, however calcium influx may not be a prerequisite for egg activation, since some parthenogenetic agents do not require the presence of external calcium (Steinhardt, Zucker and Schatten, 1977). This observation suggests that activation is accompanied by the mobilization of internal stores of bound calcium. In contrast to the situation in sea urchins, in the surf clam exogenous calcium is required for egg activation (Schuetz, 1975).

The calcium ionophore, A-23187, stimulates the release of internal calcium in echinoids, annelids, echiuroids, and tunicates and activates development (Steinhardt *et al.*, 1974). The calcium flux appears to follow the fertilization potential suggesting that the latter is a direct consequence of the interaction of the sperm and egg plasma membranes (Whitaker and Steinhardt, 1982). Using the calcium-dependent, luminescent protein, aequorin, a transient increase in calcium has been demonstrated in sea urchin, starfish, and medaka eggs (Gilkey *et al.*, 1978; Moreau *et al.*, 1978; Eisen *et al.*, 1984). The exact location of

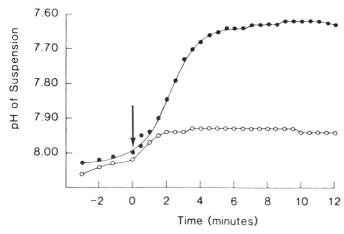

Figure 5.2 pH of sea urchin (*Arbacia*) eggs suspended in sea water (●) and sodium-free sea water (○) at various times after activation with calcium ionophore, A-23187. Reproduced with permission from Carron and Longo (1982).

released calcium has been followed in medaka ova and appears as a narrow band starting at the site of sperm contact, propagating to the opposite pole of the egg. The calcium wave precedes exocytosis of cortical granules.

In addition to the dramatic change in calcium that eggs undergo at fertilization there is an increase in intracellular pH. In sea urchin eggs the internal pH changes from 6.9 to 7.3 with a concomitant increase in acidity of the sea water; the egg remains alkaline for approximately 60 min (Johnson, Epel and Paul, 1976; Shen and Steinhardt, 1978; Fig. 5.2). There is evidence suggesting that the eggs of other organisms also undergo a change in internal pH during activation (Shen, 1983). Fertilization of starfish eggs (*Asterias*) includes the production of acid in the surrounding sea water which suggests that a rise in internal pH occurs with activation. Acid production is not universal among eggs of invertebrates, e.g. it is not seen in fertilized *Mytilus* (lamellibranch), *Acmanea* (gastropod) and *Ascidia* (tunicate) ova. Possible changes in intracellular pH in these species have not been explored.

The increase in internal pH at fertilization requires sodium, as acid production is a result of an efflux of protons from eggs and is a function of the external sodium concentration (Johnson, Epel and Paul, 1976). This also suggests that the sodium and proton fluxes are coupled. In addition, amiloride, which inhibits passive sodium flux, blocks acid production and eggs remain unactivated. The exchange of sodium and proton has a 1:1 stoichiometry that lasts for about 3 min and is not inhibited by sodium cyanide. It is normally during this period that the cortical granule reaction and other cortical modifications in sea urchin eggs take place.

In addition to sodium/proton fluxes there is a sodium/potassium exchange, detectable within minutes of insemination, that causes the sodium content of the egg to fall below that of the unfertilized egg. This exchange is energy-dependent and reduced in sea water with a low potassium concentration (Girard, Payan and Sardet, 1982). That the ratio of sodium/potassium can be minimized by lowering the external potassium concentration suggests that sodium/potassium exchange is mediated by a sodium/potassium ATPase. Investigations with sea urchins (*Paracentrotus*) show that after fertilization the alkalinity of the cytoplasm is maintained by two mechanisms: a sodium/proton exchange with the same characteristics as in unfertilized eggs, and an acid-extruding pump that is dependent on external sodium, amiloride sensitive and requires metabolic energy (Payan, Girard and Ciapa, 1983).

Although acid production and cytoplasmic alkalinization are sodium-dependent and amiloride-sensitive, a number of observations indicate that they may not be directly coupled (Shen, 1983).

1. The role of acid production is linearly dependent on external sodium, while the rate of cytoplasmic alkalinization is independent of external sodium concentrations above a minimal concentration necessary for egg activation.
2. Eggs activated with NH_4Cl and then washed in fresh sea water release acid upon fertilization. However, fertilization of NH_4Cl-pulsed eggs does not induce cytoplasmic alkalinization.
3. Acid production starts 30–60 s postinsemination and is completed by 2–4 min later. The rise in intracellular pH begins 60–90 s postinsemination and is completed by 6–8 min postinsemination.
4. The amount of acid produced varies with the external pH; at pH 9 it is greater than at pH 7.

The effects of increased intracellular pH in activated sea urchin eggs are numerous and varied (Shen, 1983; Table 5.2). Many of the pH-induced events are not interdependent, e.g. the increased potassium conductance can be suppressed without affecting protein synthesis. Activation of DNA synthesis can occur without increased protein synthesis and protein synthesis can be enhanced without DNA synthesis or chromosome condensation. The fact that these pH-induced events

Table 5.2 Processes in sea urchin eggs affected by the increase in intracellular pH at fertilization. Data compiled from Shen (1983).

Process	Effect*
Protein synthesis	↑
DNA synthesis	↑
Messenger RNA polyadenylation	↑
Glucose-6-phosphate dehydrogenase activity	↑
Pronuclear development	+
Chromatin condensation	+
Potassium conductance	↑
Glycogenolysis	↑
Thymidine uptake	↑
Cortical granule reaction	−

* ↑, increased; −, no effect; +, promoted.

are not mutually dependent on one another suggests that fertilization induces a single pervasive change, which then activates a series of independent fertilization responses (Epel, 1978). The mechanism by which the increase in pH induces such diverse responses is not known. One possibility is that cytoplasmic alkalinization affects specific enzymatic activities and protein–protein interactions which lead to regulatory modifications such as phosphorylation.

The addition of NH_4Cl (5 mM, pH 8.0) to unfertilized sea urchin

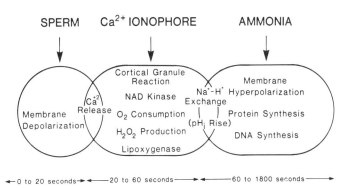

Figure 5.3 Postactivation sequence of events by sperm, calcium ionophore or ammonia. The normal sequence, initiated by sperm results in the initial membrane depolarization, calcium release and all subsequent changes. Calcium ionophore does not initiate the initial membrane depolarization, but triggers subsequent events. Incubation in ammonia or other weak bases induces the rise in intracellular pH. The earlier changes do not take place, but those changes following the intracellular pH rise are initiated. See Epel (1980).

eggs stimulates their metabolism and ability to undergo chromosome replication and condensation without triggering early surface reorganization of the egg, such as the cortical granule reaction and microvillar elongation (Epel *et al.*, 1974). Presumably ammonia does not induce a transient increase in calcium and acts by alkalinizing the cytoplasm.

In an effort to determine causal relationships between structural and metabolic changes that are initiated by fertilization or artificial activation, Epel (1980) has called attention to the fact that the sequence of fertilization/activation can be divided into two temporarily distinct series of events (Fig. 5.3). The first set of events, which includes early processes, such as the cortical granule reaction, activation of NAD kinase, lipoxygenase and an increase in oxygen consumption, is triggered by the transient rise in calcium ions. The second set which includes relatively later processes, such as membrane hyperpolarization, an increase in protein synthesis and the initiation of DNA synthesis, arise from the earlier calcium increase and are due to the transient activation of a sodium–proton exchange with a resultant rise in pH.

6

Blocks to polyspermy and the cortical granule reaction

In most animals fertilization is normally monospermic, i.e. only one sperm enters the egg and together with the maternal chromosomes forms the embryonic genome. The entry of more than one sperm, leads to abnormal development in many invertebrates, fish, anurans and mammals, and is referred to as pathologic polyspermy. In some animals, particularly those with large eggs such as urodeles, reptiles and birds, multiple sperm entry is common. However, only one male pronucleus becomes associated with the female pronucleus to constitute the embryonic genome. This situation is known as physiological polyspermy. To prevent the lethal effects of polyspermy, specific mechanisms, or blocks to polyspermy, have evolved to allow only one sperm to enter the ovum or to participate in the development of the embryonic genome. Such mechanisms include the following.

1. A rapid block, characterized as transient and incomplete, which involves the depolarization of the egg plasma membrane.
2. A slower block, usually involved with the exocytosis of cortical granules from the egg, which is complete and accomplished within minutes of insemination.
3. In some cases monospermy is ensured by specialized functions of the egg cytoplasm which are unlike processes 1 and 2.

In the eggs of many animals more than one mechanism prevents polyspermy.

The rapid, partial block to polyspermy lowers the receptivity of the egg to further sperm fusion less than 2–3 s after insemination and is associated with a depolarization of egg plasma membrane (Jaffe, 1976; Whitaker and Steinhardt, 1983). Although the existence of a rapid block to polyspermy has been questioned, it has a physiological mechanism based on the electrical response of the egg to fertilization (discussed in Chapter 5). The membrane potential of the egg depolarizes at fertilization, simultaneously with the occurrence of the rapid block to polyspermy. If the egg is subjected to a voltage clamp at a positive membrane potential, fertilization is blocked, whereas if depolarization is blocked by voltage clamping, polyspermy occurs. These observations strongly suggest that the change in the egg plasma membrane potential at fertilization regulates sperm-egg receptivity.

In sea urchins, which have been studied extensively regarding polyspermy blocks, the electrically mediated fast block to polyspermy is followed by the discharge of vesicles from the egg cortex – the cortical granule reaction. The cortical granule reaction is characteristic of eggs from many organisms including echinoderms, fish, amphibians, and mammals. In some animals such as pelecypods and some annelids, cortical granules are present but do not undergo exocytosis or change with fertilization, nor do extraneous layers develop around the egg. Nevertheless, a block to polyspermy takes place within seconds after fertilization or artificial activation in these organisms. The eggs of some organisms, e.g. the ascidian, *Ciona*, reportedly do not have cortical granules. At insemination in *Ciona*, material reportedly contained in subcortical vesicles is released from the ovum (Rosati, Monroy and de Prisco, 1977).

The cortex of the sea urchin egg is lined with a layer of cortical granules about 1 μm in diameter (Figs. 6.1, 6.2). In *Strongylocentrotus* there are about 18 000 of these organelles per egg (Vacquier, 1981). They are manufactured by the Golgi complex and become closely associated with the plasma membrane during oocyte development (Anderson, 1968). In *Ophiopholis* (an ophiuroid) the cortical granule population forms a layer five- to six-deep, and at insemination they move to and fuse with the egg surface (Holland, 1979). A similar massive exocytosis of cortical granules also occurs in the shrimp (*Penaeus*) and cnidarian (*Bunodosoma*) (Dewel and Clark, 1974; Clark *et al.*, 1980).

Although the structure of the sea urchin egg cortex has been analyzed by a number of techniques, the nature of the association of cortical granules and the plasma membrane remains an enigma (Detering *et al.*, 1977). Analyses of the sea urchin egg cortex show the cytoplasmic region associated with the cortical granules and plasma membrane to be relatively unspecialized, lacking any apparent modification which might serve to attach the two structures. However, the connection of cortical granules to the oolemma is sufficiently strong to survive forces encountered during the isolation of plasma membrane–cortical granule

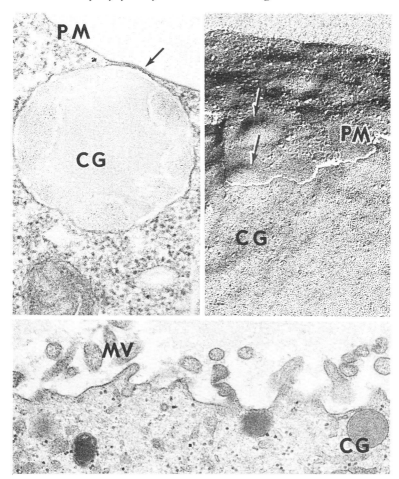

Figure 6.1 (*top left*) Cortical granule (CG) of a sea urchin (*Arbacia*) egg closely associated with the plasma membrane (PM). The arrow depicts a dome-shaped elevation where the two structures are closely associated.

Figure 6.2 (*top right*) Freeze-fracture replica of a cortical granule (CG) from a sea urchin (*Arbacia*) egg and a portion of the P-face of the plasma membrane (PM) showing dome-shape elevations (arrows) where the two structures are closely associated with one another. See Longo (1981a).

Figure 6.3 (*bottom*) Cortex of a hamster egg showing microvilli (MV) and cortical granules (CG).

complexes. The normal attachment of sea urchin eggs to the overlying plasma membrane can be disrupted by urethane and tertiary amines, suggesting that a special attachment exists between the two structures (Hylander and Summers, 1981).

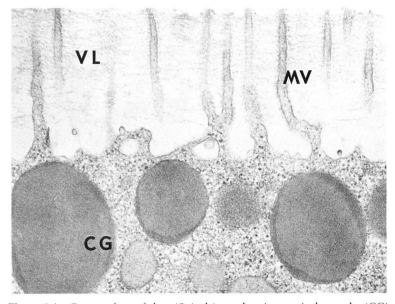

Figure 6.4 Cortex of a surf clam (*Spisula*) egg showing cortical granules (CG) and microvilli (MV). VL, vitelline layer.

Modifications of sea urchin egg plasma membranes have been observed in areas occupied by cortical granules using freeze-fracture replication, scanning electron microscopy, and filipin staining for 3-β hydroxysterol components (Longo, 1981a; Carron and Longo, 1983). The plasma membrane modifications seen with freeze-fracture replication are dome-shaped areas lacking intramembranous particles (Figs. 6.1, 6.2). These may allow specific contacts between the plasma membrane and the cortical granules thereby facilitating bilayer fusion. Unique patterns of intramembranous particles and particle free areas within the plasma membrane, and possibly induced by underlying structures, have been described in numerous cells having secretory activities. Distinct clearings of intramembranous particles have been observed in portions of the plasma membrane associated with secretory vesicles and are generally considered to represent areas depleted of membrane proteins at the fusion zone.

Ultrastructurally, the contents of sea urchin cortical granules display variations in organization depending upon the species. The content of *Arbacia* cortical granules is distinguished by a central, scalloped mass, surrounded by some lenticular material (Fig. 6.1). In *Strongylocentrotus* there is a spiral of electron-dense material which is associated with

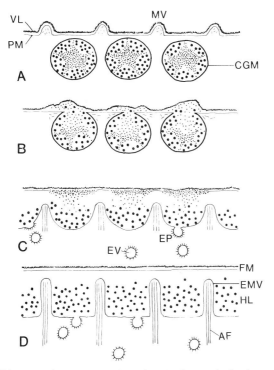

Figure 6.5   Diagramatic representation of cortical granule discharge, fertilization membrane formation, hyaline layer development, microvillar elongation and the initiation of endocytosis in sea urchin eggs. (A) Cortex of the egg depicting cortical granules, plasma membrane (PM), vitelline layer (VL) and short microvilli (MV). CGM, cortical granule membrane. (B) Cortical granule discharge and vitelline layer elevation. (C) (D) A portion of the cortical granule contents has joined with the vitelline layer to form the fertilization membrane (FM). The remaining cortical granule material remains in the perivitelline space to become the hyaline layer (HL). Portions of the plasma membrane are involved in endocytosis as evident by the presence of endocytotic pits and vesicles (EP and EV). The surface of the fertilized egg is projected into elongate microvilli (EMV) containing a core of actin filaments (AF).

some amorphous material. The complex organization of cortical granule components in other organisms has also been described, e.g. the mussel, *Mytilus* (Humphreys, 1967). The cortical granules in amphibian and mammalian eggs do not show unusually complex patterns and are filled with electron-dense granular material (Gulyas, 1980; Fig. 6.3). Cortical granule content from sea urchins, amphibians, and mammals have been examined directly by biochemical and cytochemical techniques and indirectly by analysis of the medium following their

Table 6.1 Constituents of cortical granules (see Schuel, 1978).

Component	Organism	Localization* Morphologic	Biochemical	Cytochemical	Secretory
Calcium	Sea urchin	– –	+	– –	
Protease	Sea urchin	– –	+	– –	+
	Mammal	– –	– –	– –	+
Peroxidase	Sea urchin	– –	+	+	+
Sulfated mucopolysaccharide	Sea urchin	– –	+	+	+
	Amphibian	– –	– –	+	– –
	Mammal	– –	– –	+	+
β,3-glucanase	Sea urchin	– –	+	– –	+
Hyaline protein	Sea urchin	+	– –	– –	+
	Amphibian	+	– –	+	+
Acid phosphatase	Sea urchin	– –	– –	+	– –
	Mammal	– –	– –	+	– –
β-glucuronidase	Amphibian, Reptile and Bird	– –	– –	+	– –

*– –, not present/not determined; +, present.

discharge (Schuel, 1978). Calcium, serine protease and sulfated muco-polysaccharides appear to be universal components of these structures. Peroxidase, $\beta1,3$-gluconase, hyaline protein, β-glucuronidase, and other proteins are also present in the cortical granules of some organisms (Table 6.1).

The plasma membrane of sea urchin eggs is reflected into relatively short microvilli which lack a core of actin microfilaments (Fig. 4.2). The underlying cortical granules tend to be situated in areas which lack microvilli (Schroeder, 1979). In amphibian, molluscan and mammalian eggs the microvilli are relatively longer (Fig. 6.4), and contain a microfilamentous core. Attached to the sea urchin oolemma is a glycocalyx, or vitelline layer (Anderson, 1968). It is this structure to which sperm bind via bindin and that at the time of cortical granule exocytosis, becomes detached from the egg surface to form the fertilization membrane (Fig. 6.5). The intervening space, i.e. the region between the elevated vitelline layer and the egg plasma membrane, filled with secretory materials derived from cortical granules, is the perivitelline space (Millonig, 1969; Chandler and Heuser, 1980). It is important to note that the elevation of the vitelline layer to form a fertilization membrane, as in the case of sea urchins, is not a feature common to the eggs of all animals that undergo cortical granule exocytosis at fertilization.

Studies show that the sperm initiates an exocytotic wave which traverses the egg at about 10 to 20 $\mu m\ s^{-1}$ (Jaffe, 1983). The kinetics of this wave are consistent with those of an autocatalytic process. There are observations, however, which are not entirely consistent with the notion that the cortical granule reaction is propagated by autocatalysis. For example, localized cortical granule discharges have been observed under a variety of experimental conditions (Chambers and Hinkley, 1979). These results also demonstrate that the release of cortical granules does not automatically induce the discharge of neighboring granules. Furthermore, a wave of cortical granule discharge can pass through areas lacking cortical granules, i.e. transmission of the stimulus for exocytosis does not require the presence of cortical granules.

Exocytosis of cortical granules has been studied in echinoderms, amphibians and mammals and appears to involve similar processes in these species (Figs. 6.5, 6.6). The molecular mechanisms responsible for membrane fusion and exocytosis have not been identified. Calcium may play a role in the electrostatic attraction of the cortical granules to the plasma membrane; activation of serine proteases and phospholipases may also be involved with the fusion of the cortical granule and plasma

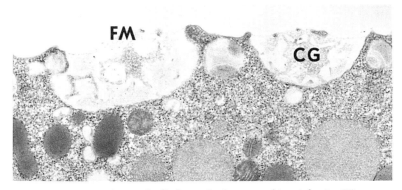

Figure 6.6   Cortical granule discharge in the sea urchin, *Arbacia*. CG, exocytosing cortical granules; FM, developing fertilization membrane.

membrane. Using transmission electron microscopy Anderson (1968) and Millonig (1969) indicated that cortical granule opening may occur via multiple fusions between the cortical granule membrane and oolemma and thereby a series of vesicles, composed of membrane derived from both the cortical granule and oolemma, are released to the perivitelline space. Whether this scheme is correct is important in establishing qualitative and quantitative changes of the egg surface area at fertilization. Using freeze-fracture replication Chandler and Heuser (1979) were unable to find intermediate stages of cortical granule membrane–plasma membrane fusion, suggesting that the fusion process is completed very rapidly. They indicated that a single pore is formed which increases in size to allow dehiscence of the cortical granule contents. This suggests that all of the membrane delimiting a cortical granule is incorporated into the egg plasma membrane when the two structures fuse.

The current paradigm regarding the mechanisms of cortical granule discharge is that calcium functions as an essential intracellular messenger. This is based on the following observations:

1. Isolated preparations of cortical granules (cortical granule lawns) can be made to exocytose by the addition of calcium (Vacquier, 1975).
2. Calcium ionophore, A-23187, triggers the cortical granule reaction in sea urchin eggs without external calcium (Steinhardt *et al.*, 1974; Chambers, Pressman and Rose, 1974).
3. Studies in the fish, medaka, are consistent with a self-propagating calcium-induced calcium release initiated by the sperm (Gilkey *et al.*, 1978).

4. The cortical granule reaction is inhibited in sea urchin eggs micro-injected with the calcium chelator, EGTA.

The threshold concentration of calcium necessary to induce the cortical granule reaction *in vitro* depends upon a number of factors including: the preparation of eggs, the ionic composition of the suspending medium, the concentration of ATP, and the EGTA-calcium binding constant used to compute the calcium concentration (Shen, 1983). Nine to 18 μM calcium resulted in cortical granule discharge from isolated cortical granule preparations (Steinhardt, Zucker and Schatten, 1977). Values of 1–3 μM affected cortical granule discharge when ATP was supplied to the medium and the cortices were prepared in solutions closely resembling the ionic composition of the egg cytoplasm (Baker and Whitaker, 1978). With isolated cortical granule preparations from the sea urchin, *Hemicentrotus*, two classes of calcium sensitivity in different media have been demonstrated (Sasaki, 1984). In *Hemicentrotus* there is also evidence for a dissociable protein which may regulate the calcium sensitivity of cortical granule discharge.

The release of calcium from different stores has been demonstrated for sea urchin eggs. A-23187 and monoelectrolytic activators release calcium from an undefined source – as occurs during fertilization. In contrast, hypertonic activating treatments appear to release calcium from a different cellular compartment. The release of calcium appears to involve an 'all or nothing' phenomenon and a repeat release is possible after about 40 min. This 40 min period is believed to represent the time required to recharge the internal store (Zucker, Steinhardt and Winkler, 1978).

The nature of the internal store of calcium is uncertain. Almost all calcium binding ability of the unfertilized sea urchin egg is found in a large particulate fraction (microsomes?) isolated by differential centrifugation (Steinhardt and Epel, 1974). Preparations of vesicles derived from *Xenopus* eggs are able to sequester calcium in an ATP-dependent manner (Cartaud, Boyer and Ozon, 1984). Studies using cytochemical reagents demonstrate the presence of calcium on the plasma membrane, cortical granules, cortical endoplasmic reticulum and other organelles of sea urchin eggs (Sardet, 1984).

Although the source(s) of internal calcium for cortical granule discharge has not been identified, specialized regions of the egg endoplasmic reticulum that are associated with the cortical granule in amphibian (*Xenopus*), sea urchin and mouse ova have been implicated (Gardiner

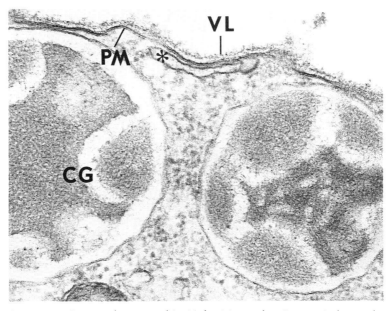

Figure 6.7   Cortex of a sea urchin (*Arbacia*) egg showing cortical granules (CG) and a cisternum of endoplasmic reticulum (*) closely associated with the egg plasma membrane (PM). VL, vitelline layer. See Luttmer and Longo (1985).

and Grey, 1983; Fig. 6.7). This suggestion is derived from observations demonstrating the following.

1. Muscle contraction is mediated by calcium release from a specialized subsurface cisternae of endoplasmic reticulum, the sarcoplasmic reticulum.
2. The striking morphological similarity of the plasma membrane–endoplasmic reticulum association observed in *Xenopus*, mouse and sea urchin eggs to the transverse tubule and sarcoplasmic reticulum of muscle cells (Gardiner and Grey, 1983; Sardet, 1984; Luttmer and Longo, 1985).
3. The temporal correlation in the development of the cortical endoplasmic reticulum and capacity of *Xenopus* eggs to propagate a wave of cortical granule exocytosis (Charbonneau and Grey, 1984; Campanella *et al.*, 1984).

In *Xenopus* eggs cortical endoplasmic reticulum cisternae are arranged parallel to the plasma membrane; the extent of opposition is 70–100 nm with a gap between the membranes of 8–13 nm. Based on these similarities it has been postulated that the close association of the plasma membrane and cortical endoplasmic reticulum in eggs transduces the interaction of gametes into intracellular calcium release that triggers the cortical granule reaction and the activation of development.

Properties of the calcium compartment in eggs have not been defined. Energy is required to fill the stores since depletion of ATP blocks the cortical granule reaction and there is an ATP-dependent binding of calcium-45 in isolated cortices (Baker and Whitaker, 1978). Consistent with these observations are investigations demonstrating a calcium-dependent ATPase in sea urchin cortices (Mabuchi, 1973). Interestingly, agents that inhibit calcium release from the sarcoplasmic reticulum, such as sodium dantrolene, have no effect on cortical granule discharge in sea urchin eggs at fertilization.

That the cortical granule reaction is characteristic of a self-propagating wave rather than diffusion suggests that the wave of calcium release is initiated by the sperm (Gilkey *et al.*, 1978). The necessary increase in free calcium sufficient to activate the medaka egg occurs as a wave of calcium-stimulated calcium release. The calcium wave is still present when the cortical granules are centrifuged to one pole of the egg suggesting that the cortical granules are not the source of calcium and another egg component is involved, possibly cortical endoplasmic reticulum.

Gamete adhesion may account for the calcium release although attachment of bindin does not induce the cortical granule reaction (Vacquier and Moy, 1977). Electron microscopic observations of early fertilization events in *Lytechinus* eggs demonstrate that the cortical granule reaction follows sperm egg fusion (Longo *et al*, 1986). However, membrane fusion *per se* does not induce the cortical granule reaction – as demonstrated by the egg–egg fusion experiments of Bennett and Mazia (1981a,b). Membrane depolarization at fertilization may trigger the calcium release; although depolarization by direct injection of current into sea urchin eggs does not bring about activation (Jaffe, 1976).

The calcium influx associated with membrane depolarization suggests that it may take place via voltage-gated channels. The sea urchin egg can be depolarized by elevated potassium which opens calcium channels that can be blocked by verapamil. Although the rate of calcium entry in this case is not equivalent to that observed at fertilization, the level achieved is the same as that of fertilized eggs, yet the ova are not activated (Schmidt, Patton and Epel, 1982). This could mean that calcium influx alone is not adequate to activate the egg. Another possiblity is that the egg in this situation sequesters entering calcium so that its free concentration never reaches the level sufficient for egg activation. Based on these data a prevailing hypothesis for activation of sea urchin eggs is that a small amount of calcium is released at fertilization which triggers additional calcium release from an internal store, possibly cortical endoplasmic reticulum. This results in a self-propagating calcium response.

How and what initiates the calcium release and cortical granule exocytosis has been the subject of considerable inquiry and speculation. Since sea urchin sperm undergo an uptake of calcium when activated they may function as a 'calcium bomb', releasing calcium when they fuse with the ovum (Jaffe, 1983). Phospholipase A_2, a calcium dependent enzyme that cleaves the fatty acid at the second acyl position of phospholipids, plays a role in the cortical granule reaction of sea urchin eggs. Unsaturated fatty acids, such as arachidonic acid, are found at the second acyl position and phospholipase A_2 activity leads to the formation of arachidonic acid and lysophosphoglyceride which may mediate fusion of the cortical granules with the egg plasma membrane (Schuel, 1978).

Several lines of evidence support the involvement of arachidonic acid and lysophosphoglyceride in the exocytosis of cortical granules (Ferguson and Shen, 1984). Phospholipase A_2 is present in the sea urchin egg and melittin, a phospholipase A_2 activator, triggers the cortical granule reaction. The cortical granule and the egg plasma membranes are similar in their fatty acid composition, both have unusually high levels of arachidonic acid. At fertilization the level of phosphatidylcholine in the egg decreases twofold; the level of arachidonic acid in phosphatidylcholine, phosphatidylinositol-phosphatidylserine and phosphatidylethanolamine also decreases. During fertilization there is a transient activation of lipoxygenase that converts free arachidonic acid to hydroxy fatty acid (HETE, hydroxyeicosatetraenoic acid) which may be important in regulating membrane permeability.

The role of lipids, particularly phosphatidylinositol, in generating intracellular signals has been demonstrated in a number of different cell types including eggs (Berridge and Irvine, 1984; Fig. 6.8). Phosphatidylinositol is hydrolysed by phospholipase C to diacylglyceride and inositol triphosphate as part of a signal transduction mechanism for controlling a variety of cellular processes including secretion, metabolism, phototransduction and cell proliferation. Diacylglycerol operates within the plane of the membrane to activate protein kinase C, whereas inositol triphosphate is released into the cytoplasm to function as a second messenger for mobilizing intracellular calcium. Experiments to

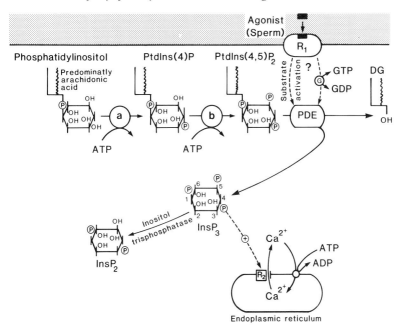

Figure 6.8 Proposed role of inositol triphosphate as an intracellular second messenger. Agonist (sperm) binds to an external receptor (R_1) to stimulate the hydrolysis of PtdIns $(4,5)P_2$ by a phosphodiesterase (PDE, phospholipase C) to form diacylglycerol (DG) and $InsP_3$. $InsP_3$ may bind to a specific receptor (R_2) on the endoplasmic reticulum to release calcium. The action of $InsP_3$ is curtailed by inositol triphosphatase which removes a phosphate from the 5-position to form $InsP_2$. a, Phosphatidylinositol kinase; b, PtdIns (4)P kinase. See Berridge and Irvine (1984).

identify the inositol-sensitive release site indicate that it is probably the endoplasmic reticulum. The induction of cortical granule dehiscence with microinjections of inositide triphosphate and the increase in polyphosphoinositide turnover in fertilized sea urchin eggs are consistent with the notion of inositol triphosphate or phospholipase C triggering the autocatalytic cycle of calcium release at fertilization (Whitaker and Irvine, 1984; Turner, Sheetz and Jaffe, 1984).

Calmodulin is associated with cortical granule-plasma membrane preparations and when isolated cortices are treated with antibody to calmodulin they lose their sensitivity to calcium (Whitaker and Steinhardt, 1982). These results and observations that calmodulin antagonists, trifluorperazine and chlorpromazine, inhibit cortical granule exocytosis are consistent with the notion that calmodulin may play

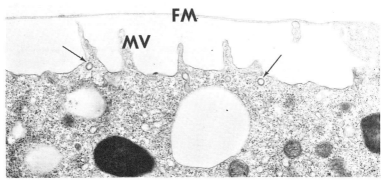

Figure 6.9   Fertilized sea urchin (*Arbacia*) egg having completed the cortical granule reaction. The fertilization membrane (FM), elongate microvilli (MV) and endocytotic vesicles (arrows) are depicted.

a role in cortical granule discharge (Baker and Whitaker, 1979; Moy *et al.*, 1983).

At the completion of the cortical granule reaction in the eggs of sea urchins, amphibians and fish, virtually all the cortical granules have been discharged. In mice a substantial number of cortical granules (about 25% of the population) are exocytosed before sperm–egg fusion; the remainder are dehisced at fertilization (Nicosia, Wolf and Inoue, 1977). In the annelid, *Sabellaria*, the cortical granule reaction is initiated when the eggs are spawned into sea water (Pasteels, 1965). Modifications in the egg cortex involving microvillar elongation take place at fertilization in *Sabellaria*. In the echiuroid worm, *Urechis*, a subset of cortical granules is released at insemination; the remainder are discharged later with the elevation of the vitelline layer (Paul, 1975).

For organisms whose eggs neither possess cortical granules nor undergo a cortical granule reaction it is clear that cortical granule exocytosis is not required for fertilization or egg activation and development. Furthermore, the cortical granule reaction can be inhibited in sea urchins and later events, including those of fertilization and cleavage, are not impaired.

As a consequence of the cortical granule reaction there is the externalization of cortical granule contents, which has profound structural and physiological effects on the egg (Figs. 6.5, 6.6). In sea urchins a portion of the cortical granule contents remains in the perivitelline space and, in the presence of calcium, polymerizes to form the hyaline layer (Citkowitz, 1971; Kane, 1973). The hyaline layer prevents polyspermy, maintains blastomere adherence and participates in morphogenetic changes of the developing embryo. Not all of the hyaline layer material

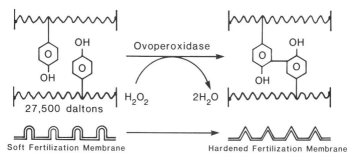

Figure 6.10 Mechanism of fertilization membrane hardening. The soft fertilization membrane contains igloo-shaped projections which represent reflections of the vitelline layer over the microvilli of the egg. The hardened fertilization membrane contains tent-shaped projections. See Shapiro (1981).

within the egg is released with the cortical granule reaction; a portion is retained in small vesicles (Hylander and Summers, 1982). These vesicles may be involved in the regeneration of the hyaline layer during embryogenesis.

The material released from the cortical granules in sea urchins also contains protease activity that promotes the elevation of the vitelline layer to form the fertilization membrane and removes sperm receptors (Vacquier *et al.*, 1973; Figs. 6.5, 6.9). Purified cortical granule contents possess two proteolytic activities involving: (1) elevation of the vitelline layer; and (2) removal of sperm receptors (Carroll and Epel, 1975). In addition other changes of the egg content are believed to be a result of proteolysis at fertilization. For example, limited proteolytic cleavage of egg surface components other than the vitelline layer occurs at fertilization and artificial activation (Shapiro, 1975). What role, if any, this limited proteolysis may serve in development has not been determined. Separation and lifting of the vitelline layer from the egg surface is accomplished by the hydration of sulfated mucopolysaccharides released from the cortical granules (Schuel *et al.*, 1974).

In the amphibian, *Xenopus*, a lectin is released from the egg during the cortical granule reaction that interacts with components of the vitelline layer in a calcium-dependent manner and alters its receptivity to sperm (Wyrick, Nishihara and Hedrick, 1974). This reaction is essentially completed within minutes of sperm entry and leads to the formation of a dense layer along the vitelline layer that, in its modified form, becomes the fertilization envelope. The fertilization envelope is relatively insoluble and functions as a barrier to further sperm incorporation.

Following its elevation, the fertilization membrane undergoes a sequence of changes resulting in a hardened glycoprotein coat that is resistant to most denaturing agents (Veron *et al.*, 1977; Shapiro, 1981). The sequence of assembly in the sea urchin, *Strongylocentrotus*, proceeds from a pliable fertilization membrane that contains casts of the microvilli, which are shaped like inverted Us, to a final hardened structure in which the microvillar casts are converted to pyramidal tent-like forms (ʌ see Fig. 6.10). During the two minutes it takes to transform from the (∩) to the (ʌ) form, the microvillar casts become coated with an orderly arrangement of repeating macromolecular units which spread to regions between these elevations (Chandler and Heuser, 1980). This coating is believed to play a role in the 'hardening' of the fertilization envelope which acts to protect the embryo. A secretory product from cortical granules has been isolated that self-associates in a calcium-dependent manner to form sheets that have a pattern similar to that seen on the fertilization membrane (Bryan, 1970a, b). Deposition of this material may be involved in hardening the fertilization membrane.

Hardening the fertilization membrane is affected by the formation of dityrosine cross-links (Foerder and Shapiro, 1977; Hall, 1978; Fig.7.10). With cortical granule exocytosis there is the release of a peroxidase (ovoperoxidase) which becomes incorporated into the fertilization membrane via a cross-linking of tyrosine residues using hydrogen peroxide from molecular oxygen. Peroxide synthesis occurs as a burst and accounts for two-thirds of the oxygen taken up by the egg during the first 15 min after insemination. Peroxide is also toxic to sperm and in addition to hardening the fertilization membrane this system may provide an additional block to polyspermy.

Polyspermy prevention in mammals may also involve physiochemical changes in the egg plasma membrane and/or the zona pellucida (the zona reaction), both of which may be mediated by cortical granule secretions. Whereas involvement of cortical granule proteins and enzymes in polyspermy prevention in sea urchins is extensively documented, the participation of a trypsin-like enzyme activity in the zona reaction and plasma membrane block in mammals is based on indirect evidence (Wolf and Hamada, 1977).

Austin and Bishop (1957) recovered eggs from the oviducts of mated mammals and quantified the incidence of multiple sperm penetration of the egg and zona pellucida. They found that blocks to polyspermy in mammals occurred at the zona pellucida and/or egg plasma membrane. In hamsters, dogs, and sheep, sperm were not seen in the perivitelline space, suggesting that the polyspermy block occurs at the level of the zona pellucida. In monospermic rabbit eggs, sperm were found in the perivitelline space, indicating that polyspermy is blocked at the plasma membrane. In mice and rats the block is believed to occur at the zona pellucida and plasma membrane. The zona reaction represents changes in the zona pellucida that (a) reduces sperm binding and penetration and (b) alters its physiochemical properties, e.g. increasing its resistance to dissolution. Sperm receptor activity has been demonstrated in solubilized preparations of zonae pellucidae from mouse eggs by competition assays (Bleil and Wassarman, 1983). The fact that zonae from two-cell mouse embryos have no receptor activity is consistent with a zona reaction.

The mechanisms responsible for the zona reaction are believed to be twofold and similar to those described for sea urchins. Release of a trypsin-like protease from cortical granules affect changes in sperm-binding properties via an alteration of sperm receptors in the zona pellucida. Peroxidase activity associated with cortical granules and with the egg surface, brings about an alteration in the zona pellucida, so making it more resistant to solubilization (Gulyas and Schmell, 1980; Schmell and Gulyas, 1980).

Evidence for cortical granule exudates effecting a block to polyspermy at the plasma membrane level in mammals is equivocal. Exposure of zonae-free eggs to crude preparations of cortical granule exudates decreases sperm penetration, which is consistent with a cortical granule involvement in a plasma membrane polyspermy block (Gwatkin, 1977). However, observations indicating that cortical granule secretions have no role in establishing a plasma membrane block have also been presented. For example, the fertility of zonae-free mouse eggs, having undergone a loss of cortical granules induced by the calcium ionophore, A-23187, was identical to that of controls (Wolf, Nicosia and Hamada, 1979). In connection with these results in mammals, sea urchin embryos removed from their fertilization membranes and hyaline layers can be refertilized, indicating the absence of a permanent, polyspermy block at the plasma membrane level in this organism as well (Longo, 1984).

In animals with large eggs, such as selachians, urodeles, reptiles and birds where physiological polyspermy is common, only one male pronucleus becomes associated with the female pronucleus. In the urodele, *Triton*, Fankhauser (1948) showed that supernumerary male pronuclei regress in the presence of the cleavage (mitotic) spindle with proximity to the spindle determining the order of regression. In mammals, pronuclear suppression reminiscent of that in *Triton* has been described in polyspermic eggs (Hunter, 1976). A process of sperm abstriction in *in vitro* zonae-free polyspermic mouse eggs has been described in which incorporated sperm are removed from the egg cytoplasm by a blebbing process that may restore the monospermic condition (Yu and Wolf, 1981).

7

Alterations in the egg cortex and cytoskeleton at fertilization

The cortical granule reaction results in a dramatic structural re-organization of the egg plasma membrane. The resultant membrane of the fertilized egg has been referred to as a mosaic, indicating that it is derived from several sources, i.e. the egg plasma membrane, the cortical granule membrane and the sperm plasmalemma. There is essentially a doubling of the surface area of the activated sea urchin egg as a result of the cortical granule reaction, i.e. the sum total of membrane delimiting all the cortical granules within the egg is equivalent to the surface area of the egg plasma membrane and both sources of membranes are believed to be combined in the cortical granule reaction (Schroeder, 1979; Vacquier, 1981). That all of the plasma membrane of the unfertilized egg and all the membrane delimiting the cortical granules are, in fact, combined to form the plasmalemma of the activated ovum has not been established unequivocally.

Formation of the mosaic plasma membrane and concomitant physiological changes in activity of the egg prompt the following questions.

1. Is the formation of the mosaic membrane related to physiological and biochemical changes characteristic of the activated ovum?
2. Do identifiable domains exist in the plasma membrane of the fertilized egg that are derived from the cortical granule or sperm plasma membranes?

It has been speculated that the contents of cortical granules serve other roles regulating the metabolism of the egg, e.g. they elicit perturbations of the plasma membrane that may be critical for activation (Schuel, 1978). In connection with this possibility it is pertinent to note that echinoid eggs can be parthenogenetically activated by trypsin and other proteases. Release of surface protein from the plasma membrane of sea urchin eggs at fertilization has been reported and it has been suggested that proteolytic processing of surface proteins may be an important aspect of activation (Shapiro, 1975).

Although activation of transport systems for specific metabolites occurs at fertilization there is no evidence linking this change with the insertion of cortical granule membrane into the plasma membrane. That transport systems develop in ammonia-activated eggs, when the cortical granule reaction is blocked, suggests that alterations in the properties induced by the cortical granule reaction may not be essential for permeability changes characteristic of fertilization (Epel, 1978). Hence, the role of the cortical granule reaction in later developmental events is uncertain.

That many aspects of fertilization are membrane-mediated events that lead to egg activation is consistent with the notion that a change in the state of the plasma membrane is an obligatory step in cellular activation. Using electron spin resonance spectroscopy Campisi and Scandella (1978, 1980a) demonstrated an increase in bulk membrane fluidity of sea urchin eggs after fertilization. However, because the spin label (fatty acid) was equilibrated among all of the subcellular membrane fractions, it could not be determined whether: (a) ovum activation is accompanied by a change in total cellular membranes to a more fluid state, or (b) more specialized membranes (such as the plasmalemma) entered a more fluid state and the probe was showing the average change experienced by altered and unaltered membranes. The structural changes of membrane lipids accompanying activation is probably not a result of the cortical granule reaction as eggs partially activated by ammonia showed a similar effect. In experiments with cortical fractions it has been shown that the fluidity of the fertilized egg cortex is less than that of the unfertilized cortex (Campisi and Scandella, 1980b). Adding calcium to cortical fractions from unfertilized eggs resulted in a fluidity decrease *in vitro*. It has been suggested that this change may represent an alteration in membrane structure rather than a direct interaction of calcium with phospholipid groups.

Analyses of membrane lipid changes in sea urchin and mouse eggs using fluorescence photobleaching recovery suggest that fertilization is not accompanied by a change in bulk membrane viscosity, rather it is associated with alterations in the ensemble of lipid domains (Wolf, Edidin and Handyside, 1981, and Wolf *et al.*, 1981). The different lipid analogs employed indicated the existence of lipid domains differing in composition or physical states from the average for the plasma membrane. These results suggest that gel and fluid lipid domains exist within the egg plasma membrane, the proportion and composition of which change upon fertilization. At fertilization there may be a reordering of lipid domains which release inactive proteins from gel regions of the plasma membrane into fluid regions, where they would become active. Changes in lipid composition and extent of gel-fluid regions at fertilization could then act as a switch which would rapidly activate protein functions that do not require the synthesis or insertion of new material into the membrane.

Studies have also been performed in sea urchin (*Arbacia*) eggs treated with filipin to detect alterations in membrane sterols at activation (see also Decker and Kinsey, 1983). The plasma membranes of treated unfertilized eggs possess numerous filipin–sterol complexes while fewer complexes are associated with membranes delimiting cortical granules, demonstrating that the plasma membrane is relatively rich in β-hydroxysterols (Carron and Longo, 1983). This dichotomy does not appear to be related to a filipin impermeability, and differences in filipin staining of the plasma and cortical granule membranes may represent an inequality in sterol content. Following fusion with the plasmalemma, membranes formerly delimiting cortical granules undergo a dramatic alteration in sterol composition – a rapid increase in the number of filipin–sterol complexes. In contrast, portions of the fertilized egg plasma membrane, derived from the original plasma membrane of the unfertilized egg, display little change in filipin–sterol composition. Other than regions involved in endocytosis, the plasma membrane of the zygote possesses a homogenous distribution of filipin–sterol complexes and appears structurally similar to that of the unfertilized ovum.

The absence of patches within the plasma membrane of the fertilized egg, which are relatively devoid of filipin–sterol complexes and which correspond morphologically to membrane formerly delimiting intact cortical granules, indicates that the cortical granule membrane is significantly altered when it fuses with the plasmalemma. How the cortical granule membrane acquires an increase in filipin–sterol complexes has not been determined. Lateral displacement of sterols from membranous regions derived from the original egg plasma membrane may be involved. However, there is no evidence documenting such a process in activated eggs. Sterols have been shown to diffuse rapidly in bilayers which is consistent with an extremely rapid lateral displacement of sterols into membranous patches derived from cortical granules.

Fluorescence photobleaching recovery experiments have been performed with mouse eggs using protein probes, suggesting that interactions with cytoskeletal components may regulate membrane protein diffusion (Wolf and Ziomek, 1983). As with membrane lipids, the proteins probed demonstrated a heterogeneous distribution. Moreover, although 'new' membranes (i.e. cortical granule and sperm plasma membrane) are added to the egg plasmalemma at fertilization, there is no generalized effect on the diffusion of membrane protein in the mouse egg.

Binding studies using plant lectins have been utilized in an effort to demonstrate possible membrane changes between fertilized and unfertilized eggs. Investigations with mouse and hamster eggs have shown that concanavalin A binding sites change quantitatively following fertilization (Yanagimachi and Nicolson, 1976). In ascidian eggs both the agglutinibility and number of concanavalin A receptors increased following activation (O'Dell, Ortolanio and Monroy, 1973). The qualitative changes in lectin binding following fertilization may reflect modifications in the nature and/or structure of the binding sites themselves. Alterations in lectin binding may also be influenced by membrane fluidity and functional states of the cytoskeleton. In the sea urchin, *Strongylocentrotus* two classes of concanavalin A binding sites have been identified: a high affinity site associated with the vitelline layer and a low affinity site associated with the plasma membrane. The number of low affinity sites doubles at fertilization, apparently as a result of the insertion of cortical granule membrane (Veron and Shapiro, 1977). Although the increase in low affinity binding sites may be due to the appearance of cryptic sites, there is no doubling when eggs are activated with ammonia. This supports the notion that the increase in number of sites is caused by the addition of cortical granule membrane to the egg plasmalemma.

Examination of freeze-fracture replicas of unfertilized sea urchin eggs demonstrates a significant difference in the number of intramembranous particles within the plasmalemma and the cortical granule membrane: in *Arbacia* the number of intramembranous particles within the P-face of the cortical granule membrane is about 30% of that in the P-face of the egg plasma membrane (Longo, 1981a). Studies have been carried out to determine what happens to this dichotomy following cortical granule exocytosis, i.e. whether there is the appearance of localized areas, corresponding to patches of cortical granule membrane, within the plasma membrane of the fertilized egg, or whether particles within the plasma membrane of the activated egg are homogeneously distributed. A homogeneous distribution of particles would suggest an intermixing of components within the mosaic membrane. The mosaic pattern of the fertilized egg plasmalemma, in terms of intramembranous particles, is temporary; recognizable differences between the original egg plasma membrane and cortical granule membrane are lost soon after cortical granule exocytosis. No patches are found which contain a reduced number of intramembranous particles and correspond to the cortical granule membrane. This indicates a rapid alteration in the composition of cortical granule membrane following its fusion with the plasma membrane. By 4 min postinsemination the density of intramembranous particles in the P-face of the plasma membrane of the fertilized egg is slightly reduced from that of the membrane of the unfertilized egg, suggesting a possible 'flow' of intramembranous particles from the oolemma into membrane derived from the cortical granules. This suggestion is in keeping with the fluid character of membranes and is consistent with schemes reported for other cell types (Frye and Edidin, 1970; Singer and Nicolson, 1972).

It has been shown that the total surface area of cortical granule membrane in sea urchin (*Strongylocentrotus*) eggs is greater than that of the egg plasmalemma (Schroeder, 1979). Hence, if all the cortical granule membrane is incorporated into the egg plasmalemma there would be at least a twofold increase in surface area. However, by 16 min postinsemination the surface area of the activated egg is only slightly larger than that of the unactivated ovum, indicating a rapid accommodation in surface membrane. The microvillar elongation that occurs following insemination may be one means of accommodating a surface increase in the activated sea urchin egg (Figs. 6.5, 7.1). However, elongated microvilli cannot compensate for all the cortical granule membrane that might be incorporated and membrane internalization has been proposed as a mechanism to quantitatively modify the surface area of activated eggs (Schroeder, 1979).

Rapid elongation of microvilli is believed to occur primarily in areas occupied by the original plasma membrane (Chandler and Heuser, 1981). Microvillar elongation may take place only at sites on the egg surface where cortical granules have exocytosed and involves a reorganization of the cortical cytoskeletal system (Vacquier, 1981).

Although actin is present in high concentration in the cortices of unfertilized sea urchin ova, few or no actin filaments are found. However, actin filaments are abundant in the cortices of fertilized eggs. These results indicate that in the unfertilized egg actin is in a monomeric form and at fertilization it polymerizes into filaments (Spudich and

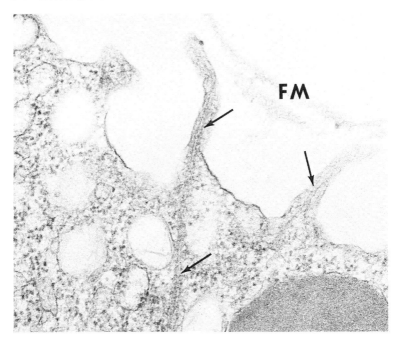

Figure 7.1   Elongate microvilli of a fertilized sea urchin (*Arbacia*) egg containing bundles of microfilaments (arrows). FM, fertilization membrane. See Carron and Longo (1982).

Spudich, 1979). In addition to actin, myosin and actin-binding proteins have been found in sea urchin eggs (Vacquier, 1981). A profilin-like protein may prevent actin from polymerizing in the unfertilized egg (Mabuchi, 1981; Hosoya, Mabuchi and Sakai, 1982).

Investigations, with both intact eggs and isolated cortices exposed to different ionic conditions, demonstrate that microvillar elongation is stimulated by the calcium flux characteristic of egg activation (Carron and Longo, 1982; Begg, Rebhun and Hyatt, 1982). Microvillar elongation does not occur when eggs are incubated in media, such as ammonia, that induce an increase in intracellular pH. However, actin filament bundle formation is triggered by an increase in intracellular pH. Formation of actin filament bundles is not necessary for microvillar elongation but is required to provide a rigid support for the microvilli. Hence, the events of activation prior to the intracellular pH increase induce the formation of cortical microfilamentous networks and microvillar elongation. The microfilaments may provide the structural and/or contractile framework for support of the egg surface which is undergoing extensive rearrangement. Microfilament organization within the microvilli, i.e. bundle formation, may then be a consequence of cytoplasmic alkalinization. Hence, actin filament bundle formation in the cortex of the sea urchin fertilized egg appears to be a two-step process: (a) the polymerization of actin filaments and (b) the association of filaments by macromolecular bridges to form bundles.

The mechanisms of cortical reorganization are not known but are likely to involve actin binding proteins as described in other systems (Pollard and Craig 1982). The distribution of α-actinin during fertilization has been investigated by microinjection of rhodamine-labeled α-actinin into living sea urchin eggs (Mabuchi *et al.*, 1985). This probe is uniformly distributed in the cytoplasm of unfertilized eggs. Upon fertilization, however, it concentrates in the zygote cortex including the fertilization cone. Migration of the fluorescently labeled egg α-actinin into microvilli apparently does not occur. Aggregation of actin filaments and their association with bundling protein e.g. fascin, may give rise to microfilament bundles in egg microvilli (Otto, Kane and Bryan, 1980). Although fascin is found in the unfertilized sea urchin egg and is localized in microvilli of fertilized ova, its interaction with actin has not been shown to be calcium- or pH-sensitive (Bryan and Kane, 1982). Hence, other actin-binding proteins may be instrumental in microvillar elongation; cytoplasmic alkalinization may give rise to microfilament bundle formation by promoting actin, actin-binding protein interactions.

In addition to changes in microvillar conformation, the eggs of a number of different animals undergo changes in cortical rigidity and contraction that appear to involve an actin–myosin system (Vacquier, 1981). Within minutes of insemination in sea urchin eggs there is an increase in cortical rigidity. In starfish eggs changes in cortical rigidity are coordinate with meiotic events. When oocytes are treated with 1-methyladenine there is a decrease in cortical stiffness. After germinal vesicle breakdown cortical rigidity remains low but increases during polar body formation. Furrow-like depressions appear along the animal vegetal axis before polar body extrusion in the oligochaete, *Tubifex* (Shimizu, 1979). Cortical contraction is prominent in fertilized frog eggs (Elinson, 1980; Kirschner *et al.*, 1980). At sperm entry there is a contraction in the animal hemisphere, called the activation wave. After the first wave ceases a second wave, the post-fertilization wave, starts at the site of sperm incorporation. Other calcium-sensitive waves follow. This contractile activity is believed to reduce the size of the animal hemisphere, thereby assisting the movements of the male pronucleus and pulling the egg surface from the forming fertilization envelope. The egg is then free to rotate within the fertilization envelope.

Cyclical changes in surface tension and contraction have been correlated with cytoskeletal alterations and also occur in anucleate egg fragments with the same cycle as in normal embryos (Yoneda, Ikeda and Washitani, 1978; Yamamoto and Yoneda, 1983). These observations indicate that egg activation initiates processes that are autonomous of the nucleus and regulate, in a cyclical manner, cytoskeletal components and cytoplasmic contraction.

Following the cortical granule reaction and concomitant with the elongation of microvilli is the development of endocytotic pits and vesicles (Donovan and Hart, 1982; Fisher and Rebhun, 1983; Carron and Longo, 1984; Sardet, 1984). Endocytosis in sea urchin eggs commences as a burst 3–5 min postinsemination in which portions of the zygote plasma membrane are taken into the cytoplasm (Fig. 6.5). Whether or not portions of the original plasmalemma or the cortical granule membrane are preferentially endocytosed has not been determined. In light of observations demonstrating significant changes in the composition of the egg plasma membrane at fertilization it seems unlikely that discrete patches of membrane persist intact to be selectively endocytosed.

That the mosaic membrane does, in fact, undergo endocytosis is demonstrated in studies employing fluid phase and absorptive tracers (e.g. horseradish peroxidase and cationized ferritin), which becomes internalized within vesicles of activated eggs. The fact that endocytosis follows the cortical granule reaction suggests a mechanism for both surface area reduction and cell surface remodeling which may be relevant to physiological changes characteristic of fertilized eggs. Its presence is consistent with observations in secretory cells where, after exocytosis, excess membrane may be removed from the cell surface. The extent of membrane internalized by endocytosis beginning at fertilization appears to be extensive and persists up to the time of cleavage. Whether endocytosis remains constant over this period has not been established, however, it has been estimated that in the sea urchin, *Strongylocentrotus*, about 26 300 μm^2 of surface membrane per egg is readsorbed by endocytosis during the first 4 min of fertilization (Fisher and Rebhun, 1983). This represents approximately 46% of the membrane presumably added to the egg surface by cortical granule exocytosis. The relationship between cortical granule exocytosis and endocytosis, in terms of the quantity of membrane in flux, is unclear since: (a) the rate of membrane interiorization is unknown; (b) the amount of cortical granule membrane added to the zygote surface has not been established and may be less than 100%; and (c) mechanisms other than endocytosis that may contribute to the reduction of surface area have not been eliminated.

Following the appearance of tracer in pinocytotic vesicles of fertilized *Arbacia* eggs, label has been observed in lysosomes (Carron and Longo, 1984). This transition indicates that the tracer travels from one cellular compartment to another. That tracer was localized to lysosomes of zygotes examined up to 60 min postinsemination also suggests that surface membrane may be degraded or modified. Membrane components may then re-enter cytoplasmic precursor pools by traversing the lysosomal membrane, to be utilized at later stages of embryogenesis.

The interiorization of zygote plasma membrane indicates the existence of a pathway from the cell surface to the cytoplasm and has significant biological implications. In a variety of cell types nutritional and regulatory molecules are selectively incorporated from the extracellular milieu. Receptor-mediated uptake and degradation systems may serve to modify hormones and surface receptors and may represent a mechanism for alteration of the physiological response of cells to their environment (Goldstein, Anderson and Brown, 1979). Whereas endocytosis may contribute to zygote surface area reduction, the interiorization of the zygote plasma membrane may participate in modifications of surface properties and may represent a mechanism for altering the number and pattern of developmentally significant cell-surface receptors.

8

Resumption of meiotic maturation

Early cytologists established much of the nomenclature and the sequence of meiotic maturation. Oocyte maturation is accomplished by two successive meiotic divisions during which the chromosome number is halved. This reduction is achieved by the formation of two polar bodies – two small cells which are formed during the meiotic divisions and serve as 'receptacles' for extra chromosomes which the egg/zygote eliminates. The haploid chromosomes that remain within the egg/zygote become a part of the female pronucleus and eventually 'one-half' of the embryonic genome. More contemporary studies of meiotic maturation have been concerned with chromosome structure and mechanisms involving the regulation of this process (Masui and Clarke, 1979).

As previously indicated, eggs of various molluscs and annelids are normally inseminated during meiotic prophase (the germinal vesicle stage) while eggs of some organisms, such as starfish, are fertilized during germinal vesicle breakdown, just prior to the formation of the first meiotic apparatus. The sequence of events in eggs fertilized at the germinal vesicle stage differs from that of ova fertilized at later stages of meiosis. In the former, this variation primarily involves germinal vesicle breakdown, formation of a meiotic spindle and its movement to the periphery of the zygote. Morphogenesis of the maternally derived chromatin of eggs belonging to classes I–III (Fig. 4.10), involving the later stages of meiotic maturation, e.g. polar body formation, are comparable in all other aspects. Figure 8.1 illustrates some of the major chromosomal events occurring during meiosis.

One of the most dramatic indications of activation in eggs fertilized at meiotic prophase is the breakdown and disappearance of the germinal vesicle. The germinal vesicle is a large, euchromatic spheroid nucleus that often contains a large nucleolus with several distinct regions (Figs.

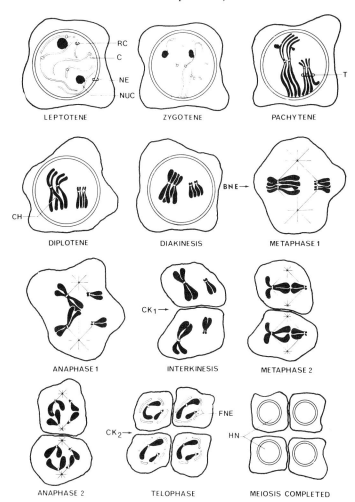

Figure 8.1 Structures and stages of meiosis. RC, replicated leptotene chromosomes; C, centromere; NE, nuclear envelope; NUC, nucleolus; T, tetrad; BNE, breakdown of nuclear envelope; CK₁ and CK₂, cytokinesis 1 and 2; FNE, formation of nuclear envelope; HN, haploid nuclei. Reproduced with permission from Longo and Anderson (1974).

8.2, 8.3). The nuclear envelope is distinguished by its smooth contour and numerous pores. Granular aggregations have been observed within the germinal vesicle in a number of different species. The composition of these structures has not been determined but they are thought to contain RNA, possibly in transit to the cytoplasm.

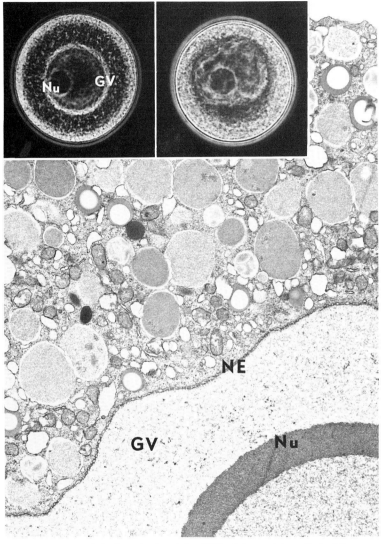

Figure 8.2 (*inset left*) Germinal vesicles (GV) of the surf clam (*Spisula*) oocyte. Nu, nucleolus.

Figure 8.3 Germinal vesicle (GV) of the starfish (*Asterias*) oocyte. NE, nuclear envelope; Nu, nucleolus.

Figure 8.4 (*inset right*) Germinal vesicle breakdown in a surf clam (*Spisula*) zygote. See Longo and Anderson (1970a).

Breakdown of the germinal vesicle is a fairly rapid event; in the surf clam, *Spisula*, it occurs 10–15 min after insemination at 20°C (Chen and Longo, 1983). This process has been examined in a number of different organisms using electron microscopy and appears to be morphologically similar in each organism examined (Calarco, Donahue and Szollosi, 1972). Initiation of germinal vesicle breakdown is recognized when its surface becomes highly irregular and plicated (Fig. 8.4). Internally the chromosomes condense and the nucleolus disappears. Subsequently, openings are seen along the nuclear envelope, owing to multiple fusions between the inner and the outer membranes of the nuclear envelope. These breaks lead to the formation of cisternae which may outline the chromosomes for a brief period but then are scattered into the cytoplasm. In some instances, e.g. human oocytes matured *in vitro*, cisternae derived from the nuclear envelope outline the condensing chromosomes as they are arranged along the metaphase plate; later the cisternae disappear.

Organelles identified as lysosomes aggregate along the germinal vesicle of rat oocytes undergoing meiotic maturation (Ezzell and Szego, 1979). It has been suggested that this mobilization of lysosomes is a specific response to factors promoting the resumption of meiotic maturaton that leads to germinal vesicle breakdown.

Meiotic maturation in mouse oocytes can be reversibly blocked at discrete stages prior to metaphase II when they are cultured in the presence of various drugs (Wassarman, Josefowicz and Letourneau, 1976). Germinal vesicle breakdown does not take place in the presence of dibutyryl cAMP; chromosome condensation is initiated but then aborts. Similarly, chloroquine, an inhibitor of lysosomal function, blocks germinal vesicle breakdown. These findings suggest that germinal vesicle breakdown may occur by a cAMP controlled protease activated mechanism, analogous to that proposed for a variety of polypeptide hormones mediating biological phenomena.

Concomitant with the breakdown of the germinal vesicle, asters appear in the egg/zygote cytoplasm. Early cytologists recognized that fertilized eggs may be initially associated with the formation of one or two asters. More recent studies with fluorescently labeled tubulin antibody have demonstrated the presence of an aster or centropshere in association with the germinal vesicle of starfish eggs (Schroeder and Otto, 1984; Otto and Schroeder, 1984). In surf clam (*Spisula*) zygotes and starfish (*Asterias*) eggs, two asters become situated on either side of the disrupting germinal vesicle. In the center of each aster in *Spisula* zygotes is located a pair of centrioles surrounded by some amorphous material. In mammalian eggs/zygotes, however, centrioles have not been observed in the asters of meiotic and cleavage spindles (Zamboni, 1971; Szollosi, Calarco and Donahue, 1972). Dense accumulations of fine textured material and vesicles are observed at the polar regions of the meiotic spindle that are reminiscent of material observed in microtubule organizing centers of somatic cells (McIntosh, 1983). The amorphous material is believed to be involved in the nucleation, polymerization, and organization of the microtubules that are a part of the asters. At the time of germinal vesicle breakdown in mouse oocytes several foci, composed of electron-dense fibrillar material from which microtubules radiate, appear near the nucleus. These aggregates form small asters and become situated at the poles of the spindle during the meiotic divisions. Fascicles of microtubules project from the centrospheres, the central portion of the asters, and are separated by areas filled with yolk bodies, mitochondria, and cisternae of endoplasmic reticulum. Microtubules appear within breaks of the nuclear envelope, become associated with condensing chromosomes and orient to the poles of the developing spindle. Concomitantly, the two asters move opposite one another and their centers define the poles of the meiotic spindle (Longo and Anderson, 1970a).

The tubulin content of surf clam (*Spisula*) eggs has been shown to be slightly greater than 3% of the total protein of the ovum (Burnside, Kozak and Kafatos, 1973). Since unfertilized *Spisula* eggs lack morphologically distinguishable microtubules, tubulin is apparently 'stored' until activation, when it is presumably utilized in the formation of the meiotic spindle. Isolated tubulin from *Spisula* eggs is capable of assembling into microtubules *in vitro* (Weisenberg and Rosenfeld, 1975). In contrast to unfertilized ova, activated *Spisula* eggs contain centrioles and granules that are capable of organizing microtubules into asters *in vitro*.

In the oocytes/zygotes of many organisms the meiotic spindle as it develops, moves to the cortex – the animal pole. In most invertebrates the spindle is positioned with its long axis perpendicular to the surface of the ovum/zygote (Fig. 8.5); in some mammals, however, the spindle is oriented tangential to the egg/zygote surface (Figs. 8.6, 8.7). Placement of the meiotic apparatus within the periphery of invertebrate eggs frequently results in an asymmetry in its structure. The aster located in

Figure 8.5 (*top left*) Unfertilized mussel (*Mytilus*) egg at the first metaphase of meiosis. The long axis of the meiotic spindle (MS) is orientated perpendicular to the egg surface. Ch, chromosomes. See Longo and Anderson (1969a).

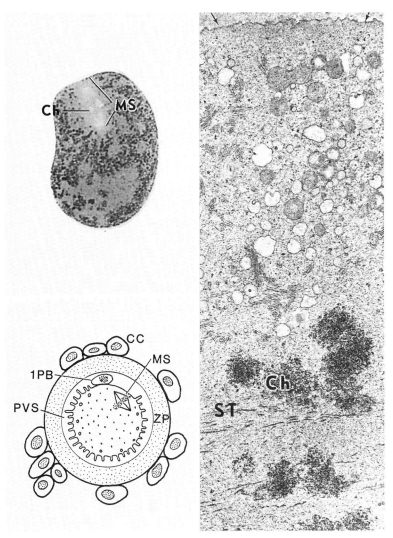

Figure 8.6 (*bottom left*) Diagramatic representation of an unfertilized mouse egg at the second metaphase of meiosis. The long axis of the meiotic spindle (MS) is oriented tangential to the egg surface. CC, cumulus cells; 1PB, first polar body; PVS perivitelline space; ZP, zona pellucida.

Figure 8.7 (*right*) Cortical region of a mouse egg containing a portion of the meiotic spindle. The surface of the egg associated with the meiotic spindle lacks microvilli immediately subjacent to the plasma membrane (arrows) is a layer of actin filaments. Ch, chromosomes; ST spindle microtubules. Taken from Longo and Chen (1985).

the cortex of the egg (the peripheral aster) is reduced in size in comparison to the centrally located aster.

In many mammalian eggs (e.g. mouse, hamster and rat) the region which overlies the meiotic spindle is distinguished by the absence of microvilli and the presence of a dense layer of actin filaments (Nicosia, Wolf and Inoue, 1977; Longo and Chen, 1985). This specialized cortical region in mouse ova is referred to as the microvillus-free area (Figs. 8.6, 8.7). Although a meiotic spindle is not formed in mouse oocytes treated with colchicine, the chromosomes move to the egg cortex and a microvillus-free area forms in the region of the cortex associated with the chromosomes. Moreover, when the meiotic spindle or the chromosomes are prevented from moving to the egg cortex a microvillus-free area does not develop. These observations suggest that interaction of the meiotic chromosomes brings about the formation of the microvillus-free area.

Subsequent to its migration to the egg cortex the meiotic spindle becomes anchored to the plasma membrane. In contrast to observations of some invertebrate eggs, cytochalasin B prevents the localization of the meiotic spindle to the cortex of maturing mouse oocytes, suggesting that a cytyochalasin B-sensitive component of the cytoskeletal system is involved in this movement (Longo and Chen, 1985). Actin has been demonstrated in nuclei and the meiotic spindle and has been implicated in force production of chromosome movements during mitosis (Zimmerman and Forer, 1981). In light of these investigations and studies demonstrating the disruptive effects of cytochalasin B on actin, an actin based system may be responsible for the cortical localization of the meiotic spindle in mouse oocytes.

Germinal vesicle breakdown is not inhibited by anaerobic conditions or uncouplers of phosphorylation in molluscs. Such conditions and agents prevent the formation of the meiotic apparatus. Spontaneous maturation of rat oocytes can be prevented by hypoxia. Germinal vesicle breakdown and chromosome condensation take place in mouse oocytes in the presence of protein synthesis inhibitors such as puromycin, however, nuclear progression is blocked at the circular bivalent stage when oocytes are cultured continuously in the presence of the drug (Clarke and Masui, 1983).

In general, changes in protein synthesis activity occur during meiotic maturation in all species studied thus far (Masui and Clarke, 1979). An increase in protein synthesis occurs during amphibian oocyte maturation (Shih *et al.*, 1978). Removal of the germinal vesicle from *Rana pipiens* oocytes affects neither the rate nor the pattern of protein synthesis. Protein synthesis inhibitors block germinal vesicle break-down and oocyte maturation in amphibians and fish; however, a substantial reduction in the level of protein synthesis occurs under anaerobic conditions that does not prevent oocyte maturation provid-ing the eggs have undergone germinal vesicle breakdown (Smith and Ecker, 1970). With inhibition of protein synthesis in starfish (*Asterias* and *Piasterias*) oocytes, germinal vesicle breakdown occurs and maturation is not arrested until metaphase I. Fusidic acid blocks protein synthesis and stops maturation before germinal vesicle breakdown in *Chaetopterus* (annelid) oocytes, whereas puromycin inhibits protein synthesis by 50% but does not affect maturation, implying that some of the protein synthesized during maturation in this organism is actually required for this process. Fertilized *Spisula* (surfclam) eggs have been shown to synthesize a cell cycle-related protein, cyclin A, which induces the resumption of meiotic maturation (Swenson, Farrell and Ruder-man 1986).

Further experiments have examined what portion of the protein synthesized by the maturing oocyte is responsible for the progression of maturation and when the proteins necessary for each step of maturation are synthesized (Masui and Clarke, 1979). Results of these studies suggest that certain protein synthesis stimulating factors, that appear in the oocyte cytoplasm during the initial period of maturation, are stored in the germinal vesicle and are released during its breakdown. Oocyte maturation is not highly dependent on RNA synthesis and DNA synthesis is not required during maturation.

In most animals oocyte maturation is dependent upon ovarian functions. Generally, only fully grown oocytes mature in response to ovarian stimuli and cease development until inseminated. The systemic factors controlling ovarian egg maturation are gonadotropins which in turn stimulate the synthesis and release of hormones that are directly active on the egg (Masui and Clarke, 1979). In only a few groups of animals is there direct evidence of a chemically defined substance acting as a maturation-inducing substance. In amphibians, progesterone appears to be the natural maturation-inducing substance, although other steroids are known to be effective. In starfish, 1-methyladenine has been shown to be the natural maturation-inducing substance (Kanatani *et al.*, 1969). Considerable controversy exists as to the precise role of gonadotropins, steroids, prostaglandins, cyclic nuc-leotides and inhibin during oocyte maturation in mammals (Channing *et al.*, 1982; Eppig and Downs, 1984).

In some animals, e.g. the surf clam (*Spisula*), the induction of oocyte maturation is known to be dependent on the presence of external calcium (Masui and Clarke, 1979). The oocytes of some marine invertebrates (the starfish, *Asterias*), however, can be induced to mature in sea water lacking calcium. Chelation of internal free calcium from oocytes by injection of EGTA inhibits maturation, whether or not maturation is dependent upon the presence of calcium in the external medium. These observations indicate that internal calcium may play a role in the initiation of oocyte maturation and are supported by experiments in amphibians (*Xenopus*) and echiuroid worms (*Urechis*) which demonstrate the release of calcium into the external medium when oocytes are stimulated to mature (O'Connor, Robinson and Smith, 1977; Johnston and Paul, 1977).

Consistent with an active role for calcium in the initiation of oocyte maturation are results of experiments employing agents known to interfere with its physiological action. Agents such as D-600, which block calcium channels, and procaine, inhibit 1-methyladenine-induced maturation in starfish oocytes as well as suppress the luminescent discharge from 1-methyladenine-stimulated, aequorin-injected oocytes (Moreau *et al.*, 1978). In the mollusc, *Barnea*, there is a calcium dependent D-600 sensitive period that lasts 3-4 min postactivation which corresponds to the period in which there is an influx of calcium. A-23187, an ionophore which facilitates the transport of calcium across membranes, induces oocyte maturation in a wide variety of animals if external calcium levels are appropriate (Steinhardt *et al.*, 1974).

One possible mechanism for calcium action at maturation is that it binds to calmodulin. Injection of calmodulin triggers meiosis reinitiation in amphibians (Maller and Krebs, 1980). Although amphibian calmodulin stimulates bovine brain phosphodiesterase, it has no apparent effect on amphibian oocyte phosphodiesterase – a critical step in the regulation of oocyte maturation (Masui and Clarke, 1979). Additional problems with this model include observations that calmodulin antagonists, chlorpromazine or fluphenazine, trigger maturation.

Sodium and potassium ion permeability and electrophysiological properties of maturing oocytes have been shown to change during maturation. These membrane alterations are necessary steps in the initiation of this process by maturation inducing substances. Specific investigations of these properties are reviewed by Masui and Clarke (1979).

Reinitiation of meiotic maturation may also be accompanied by an increase in intracellular pH (Lee and Steinhardt, 1981). The role of internal pH changes in the maturation process has not been fully elucidated. Since maturation still occurs when the internal pH of progesterone treated amphibian oocytes is constant it has been suggested that a threshold value of internal pH does not exist and that internal pH change during maturation is of little importance as a regulating parameter (Cicirelli, Robinson and Smith, 1983). Similarly, starfish oocytes do not appear to require a significant increase in internal pH in order to mature in response to 1-methyladenine (Johnson and Epel, 1982).

It has been shown for amphibian and starfish oocytes that a maturation inducing substance is effective only when applied to the ooycte surface (Masui and Clarke, 1979). The change in membrane potential of oocytes exposed to a maturation inducing substance is consistent with the proposition that maturation is initiated via the plasma membrane. There is evidence that the signal for maturation is also transmitted through the cytoplasm. For example, in fish and amphibians germinal vesicle breakdown is delayed after appropriate stimulation when the germinal vesicle is moved from the animal to the vegetal hemisphere. Gurdon (1967) demonstrated that the effect of maturation inducing substance on the germinal vesicle is indirect and showed that gonadotropins fail to induce germinal vesicle breakdown when they are injected directly into amphibian oocytes. One-methyladenine has also been shown to act on the plasma membranes of starfish oocytes to trigger meiotic resumption (Doreé and Guerrier, 1975).

The cytoplasm from enucleate, 1-methyladenine-treated starfish oocytes induces germinal vesicle breakdown when injected into untreated oocytes (Kishimoto, Hirai and Kanatani, 1981). This indicates that maturation promoting factor can be produced in the absence of the germinal vesicle, however, less material is produced than from germinal vesicle intact eggs. The cytoplasm of progesterone treated amphibian oocytes can also induce germinal vesicle breakdown when injected into untreated oocytes. Furthermore, germinal vesicle breakdown activity is present in the cytoplasm of enucleate oocyte treated with progesterone (Masui and Markert, 1971). These observations have given rise to the hypothesis that a cytoplasmic substance, responsible for promoting maturation and referred to as maturation promoting factor, appears in oocytes as a result of surface stimulation by maturation inducing substance and is a product of cytoplasmic activities not requiring genomic function (Masui and Clarke, 1979).

In amphibian and starfish oocytes maturation promoting factor attains maximum activity at germinal vesicle breakdown; its activity is reduced during pronuclear formation. The factor has been found to develop in oocytes induced to mature by serial injections of cytoplasm from maturing oocytes. Based on these observations it has been suggested that the active substance is produced by autocatalytic amplification. Germinal vesicle breakdown in amphibian oocytes, induced by injection of cytoplasm containing maturation promoting factor, is not blocked by protein synthesis inhibitors. However, since maturation is inhibited by protein synthesis inhibitors it is likely that a process preceeding the appearance of maturation promoting factor is one requiring protein synthesis.

Purification of maturation promoting factor selects for a system containing cAMP-dependent protein kinase, protein substrate and phosphatases (Wu and Gerhart, 1980). The rise in protein phosphorylation in amphibian and starfish oocytes following treatment with maturation inducing substance suggests a close relation between protein phosphorylation and maturation (Maller, Wu and Gerhart, 1977). When amphibian (*Xenopus*) oocytes are injected with maturation promoting factor, a burst of protein phosphorylation occurs immediately, followed by germinal vesicle breakdown, even in the presence of cycloheximide. These results suggest that the protein phosphorylation burst is a necessary step in the mechanism by which maturation promoting factor induces germinal vesicle breakdown. In *Rana pipiens* cAMP and cGMP levels decrease to about one-half after hormone treatment and then increase shortly before germinal vesicle breakdown. The cAMP level decreases after germinal vesicle breakdown, reaches a minimum at the time of polar body formation, and returns to the level of the ovarian oocyte when metaphase II is attained. The decrease in cAMP following oocyte stimulation may be necessary for initiation of maturation as a similar process occurs in mammals (Masui and Clarke, 1979). The role of cAMP in these cases is believed to be primarily with the regulation of protein phosphokinase activities (Maller and Krebs, 1977).

Based on earlier hypotheses (Wasserman and Masui, 1975; Maller and Krebs, 1977), Masui and Clark (1979) have proposed the following scheme to explain the regulation of oocyte maturation.

1. Maturation inducing substance, acting on the oocyte surface, induces the release of calcium.
2. Released calcium acts on a calcium-dependent regulatory protein that in turn modulates the activity of phosphodiesterase.
3. Activated phosphodiesterase degrades cyclic AMP, thus lowering the level of cyclic AMP.

4. The decrease in cyclic AMP results in the inactivation of cyclic AMP-dependent protein kinase, which is responsible for phosphorylating a presumptive protein, protein A.

5. Hence, phosphorylation of protein A, a protein constantly synthesized and degraded in its phosphorylated form, is terminated.

6. Unphosphorylated protein A, which is not subject to rapid degradation, is able then to induce the phosphorylation of maturation promoting factor precursor. This results in the formation of active maturation promoting factor.

7. Active maturation promoting factor autocatalytically phosphorylates pecursor maturation promoting factor, thus amplifying its own activity.

8. Maturation promoting factor acts to induce germinal vesicle breakdown. How maturation promoting factor induces germinal vesicle breakdown has not been determined.

The fact that the ova of most animals are arrested in some stage of meiosis at insemination suggests that eggs may accumulate an inhibitory factor during development and when fertilization occurs a mechanism releases the block to maturation. Supporting evidence for this idea comes from experiments in which cytoplasm from unfertilized amphibian (*Rana pipiens*) eggs (arrested at metaphase II of meiosis) injected into one of the blastomeres of a two-cell stage embryo arrests mitosis in the injected cell (Masui and Markert, 1971). The presumptive inhibitory agent here is referred to as cytostatic factor.

Cytostatic factor develops in amphibian (*Xenopus*) eggs following germinal vesicle breakdown *in vitro*, however, little activity can be demonstrated in ovulated eggs (Masui and Clarke, 1979). This apparent discrepancy may be due to the loss of cytostatic factor activity in naturally ovulated *Xenopus* eggs, possibly caused by a release of calcium by the egg cytoplasm due to human chorionic gonadotropin induced ovulation. Cytostatic factor-like activity has also been observed in mammalian oocytes (Balakier and Czolowska, 1977).

The process of egg activation in amphibians appears to involve the inactivation of cytostatic factor; injection of fertilized egg extracts does not affect recipients when the injection takes place within 45 s following insemination (Meyerhof and Masui, 1977). The ability of activated oocytes to inactivate cytostatic factor develops in the cytoplasm without participation of the nucleus. The sensitivity of cytostatic factor to calcium and its inactivation are believed to be linked to the calcium flux that accompanies egg activation. Metaphase arrest of meiosis is a widespread phenomenon in the oocytes of many species and whether it

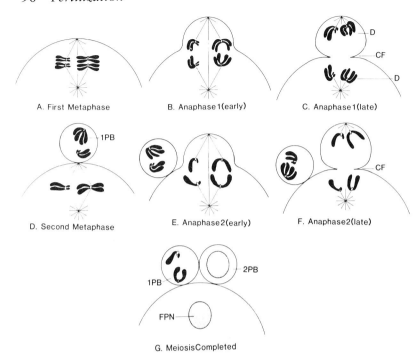

A. First Metaphase

B. Anaphase 1(early)

C. Anaphase 1(late)

D. Second Metaphase

E. Anaphase2(early)

F. Anaphase2(late)

G. MeiosisCompleted

Figure 8.8   Events involved in polar body formation. 1PB and 2PB, first and second polar bodies; FPN, female pronucleus; CF, cleavage furrow; D, dyad chromosomes. See Longo and Anderson (1974).

is due exclusively to the presence of a cytostatic factor in all organisms has not been determined.

Formation of the first polar body

Formation of the first polar body has been described in eggs and zygotes of a number of different organisms and involves the elongation of the spindle and movement of dyad chromosomes to their respective poles (Zamboni, 1971; Longo, 1973). Concomitant with these events is the production of a cytoplasmic protrusion at the animal pole which becomes the first polar body (Fig. 8.8). At anaphase I, the chromosomes become localized within this protrusion (Fig. 8.9). At the base of the protrusion a cleavage furrow is formed. The cleavage furrow associated with the formation of the first polar body in mouse eggs develops paratangentially following the rotation of the equatorial plate. Eventually the projection containing the peripheral aster and chromosomes becomes separated from the zygote. Separation is mediated by the

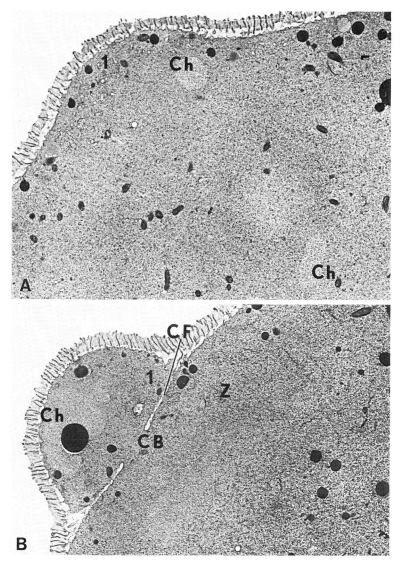

Figure 8.9 Early (A) and late (B) stages in the development of the first polar body in the mussel, *Mytilus*. 1, developing first polar body; Ch, chromosomes; CB, cytoplasmic bridge joining the first polar body and the zygote (Z); CF, cleavage furrow. See Longo and Anderson (1969a).

formation of a ring of microfilaments that is located at the base of the projection. In the case of mammalian eggs (human and mouse) a dense layer of microfilaments is also located at the distal end of the projection that becomes the polar body (Lopata *et al.*, 1980). By immunofluorescent procedures this layer, as well as the cleavage furrow, stains intensely for actin (Maro *et al.*, 1984).

The morphology of the cleavage furrow during polar body formation is similar to that developed during cytokinesis of mitotic cells (Schroeder, 1975). The electron-opaque material which lines the furrow during polar body formation is actin and may function in the manner of a 'purse string' for partitioning the cytoplasm. Progression of the cleavage furrow eventually yields a short-lived cytoplasmic bridge which joins the egg/zygote and the first polar body. A causal role for actin in polar body formation has been demonstrated in surf clam (*Spisula*) and mouse zygotes where furrow constriction and polar body extrusion are prevented by cytochalasins (Longo, 1972; Maro *et al.*, 1984). In *Spisula* inhibition of polar body formation with cytochalasin B occurs at anaphase I or II with similar results. The chromosomes that would normally be emitted are retained, thereby altering zygote ploidy. When *Spisula* zygotes are incubated in cytochalasin B throughout the course of maturation, all the chromosomes normally emitted with the polar bodies remain within the zygote and become organized into a variable number of pronuclei.

The first polar body appears as a miniature cell and has been observed in a variety of invertebrates and mammals (Fig. 8.10). A distinguishing feature of the first polar body is the presence of compacted chromatin which is usually not associated with a nuclear envelope. In many of the organisms studied, particularly mammals, the first polar body may contain cortical granules and variable amounts of endoplasmic reticulum and mitochondria (Zamboni, 1971). Large quantitites of cytoplasmic inclusions are usually found in the first polar body of the mussel, *Mytilus*, while lesser amounts are observed in the surf clam, *Spisula*. Centrioles and Golgi complexes have been observed in the first polar body of *Mytilus*. As to the possible cleavage of the first polar body, a human pronucleate zygote has been observed to be associated with three polar bodies, two of which were interpreted as being division products of the first (Zamboni *et al.*, 1966). In *Spisula* and *Mytilus*, the first polar body becomes highly electron opaque during later stages of development. This increase in electron opacity may be an indication of necrosis. The first polar body of some mammals, e.g. the rat, disintegrates soon after its formation.

Formation of the second polar body

Following the formation of the first polar body, the chromosomes remaining in the egg/zygote become aligned on the equatorial plate of

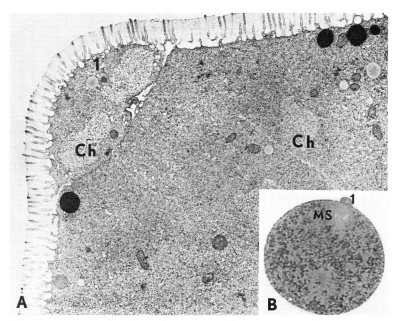

Figure 8.10 (A) and (B) Second meiotic spindle (MS) at metaphase II and first polar body of the surf clam, *Spisula*. Ch, chromosomes; 1, first polar body. See Longo and Anderson (1970a).

the second meiotic apparatus (Fig. 8.8, 8.10). This usually takes place without intervening telophase and prophase stages. The meiotic apparatus formed is structurally similar to the first and occupies an area relatively devoid of large cytoplasmic constituents (Longo, 1973). Centrioles have been found in the asters of the second meiotic apparatus in mollusc (*Spisula* and *Mytilus*) zygotes but have not been observed in mammals. In the oligochaete, *Tubifex*, the second meiotic apparatus is formed and positioned perpendicular to the egg surface 40 min after the formation of the first. The meiotic apparatus appears to be tethered to the egg surface by structural connections between filamentous elements in the cortex and microtubules of the peripheral aster (Shimizu, 1981a).

Movement of the chromosomes and elongation of the spindle at anaphase II are similar to events occurring during anaphase I (Fig. 8.11). The chromosomes move into a protrusion, approximately as wide as the peripheral aster, which is formed during spindle elongation (Longo, 1973; Shimizu, 1981b). At the base of the protrusion, a cleavage furrow develops which is morphologically similar to that observed during the formation of the first polar body; it too is

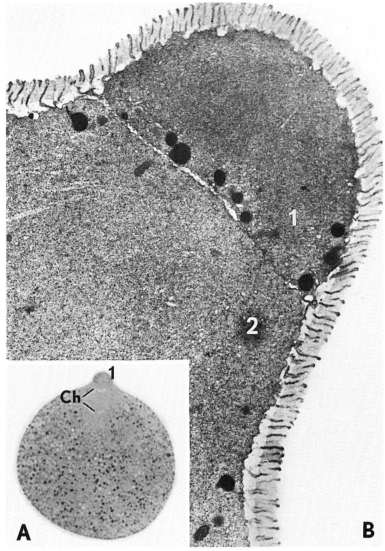

Figure 8.11 Forming second polar body (2) in the mussel, *Mytilus*. 1, first polar body; Ch, chromosomes at anaphase II of meiosis. See Longo and Anderson (1969a).

associated with a band of filamentous material along its leading edge. Progression of the cleavage furrow at the base of the developing second polar body yields a cytoplasmic bridge containing a midbody. The cytoplasmic bridge connecting the zygote and the second polar body remains at least until the first cleavage division of the zygote.

The second polar body contains a variable number of cytoplasmic constituents. Very few structures, such as mitochondria, endoplasmic reticulum, lipid droplets and yolk bodies, are observed in the surf clam, *Spisula*. In mammals the second polar body contains relatively more endoplasmic reticulum, mitochondria and yolk bodies than observed in *Spisula*. Few or no cortical granules are observed in the second polar body of the rat, hamster and human (Zamboni *et al.*, 1966; Lopata *et al.*, 1980). This is consistent with the formation of the second polar body following the release of the cortical granules in these species. The same kinds of cytoplasmic organelles and inclusions are observed in the second body of the mussel, *Mytilus*, as in the first.

The chromatin localized within the developing second polar body is initially condensed. Later, it disperses and concomitantly vesicles aggregate along its margin, fuse and form a nuclear envelope (Fig. 8.12). Chromatin delimited by a continuous nuclear envelope has been observed in the second polar body of invertebrates and mammals (Longo, 1973). Apart from its reduced size and the absence of a male pronucleus, the second polar body is structurally similar to the zygote.

Structural differences in the chromatin of the first and second polar bodies have been attributed to the lack of organelles that normally participate in nuclear envelope formation and the size difference of the zygote and polar bodies. However, the polar bodies in many species appear to contain the same kinds and number of membranous structures, and their sizes in many instances are essentially the same. Chromatin differences between the first and second polar bodies may depend upon the sequential appearance of substances regulating chromatin morphogenesis during the course of polar body formation. For example, if the egg/zygote develops the property for nuclear envelope formation and chromsome condensation after the first meiotic division, only the second polar body would be capable of forming a nucleus. Other than acting as receptacles for redundant chromatin, additional roles, if there are any, for the polar bodies during fertilization and embryogenesis have not been established.

Polar body formation may be viewed as an extreme case of unequal cleavage and represents one of the few instances where the plane of cleavage does not bisect the metaphase spindle midway between its poles (Rappaport, 1971). The cleavage furrow, which separates the polar body from the zygote, forms at the base of the prospective polar body and not at a plane in register with the metaphase plate. Asym-

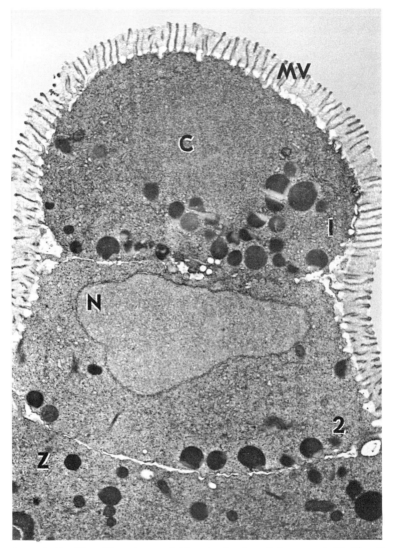

Figure 8.12 First (1) and second (2) polar bodies of the mussel, *Mytilus*. C, chromatin of first polar body; N, nucleus of second polar body; Z, zygote cytoplasm; MV, microvilli. See Longo (1983).

metrical cleavage has been shown to be associated with an inequality of aster size, for the pole containing the smaller aster is the one at which the smaller blastomere is formed (Dan and Nakajima, 1965). A similar relation may also be involved during the formation of the polar bodies.

Unequal aster size is characteristic of the meiotic apparatus in the molluscs, *Spisula* and *Mytilus* (Longo, 1973). The peripheral aster appears restricted in dimension due to its proximity to the plasma membrane. A similar situation has also been observed in dividing zygotes of *Spisula* where one aster, located in that region of the cell which becomes the smaller blastomere, is flattened on its polar surface – as if it were being pushed or pulled toward the cell's surface.

Centrifugation studies of molluscan eggs have indicated that the inequality of the meiotic division is not due exclusively to an inherent property of the maturation spindle but to the orientation of the meiotic apparatus within the egg (Raven, 1966). Centrifugation of gastropod (*Lymnaea*) eggs prior to the first and second meiotic divisions results in the formation of large polar bodies due to the movement of the meiotic spindle from the egg cortex. Normal embryonic development of the zygote occurs only when the giant polar body contains less than 25% of the egg volume.

Characteristic features of polar body formation, i.e. production of a cytoplasmic projection and the failure of the cleavage furrow to bifurcate the metaphase plate, may be due to the aster's ability to affect the cell cortex adjacent to it, causing it to become distensible. Internal pressure might then induce the cytoplasm along this region to evaginate, thereby forming a cytoplasmic mass which develops into a polar body. Before polar body formation in starfish eggs the cell surface expands at the animal pole and contracts at the vegetal pole. There is an accompanying movement of endoplasm from the vegetal pole to the animal pole (Hamaguchi and Hiramoto, 1978). With the formation of the polar body, the cell surface around the extruded region contracts and the surface at the vegetal pole expands with a movement of cytoplasm from the animal pole to the vegetal pole. Following polar body extrusion, the cell surface at the animal pole expands with a movement of cytoplasm towards the animal pole.

Development of the female pronucleus

Following anaphase II, the chromosomes remaining in the zygote disperse. Concomitantly, vesicles aggregate along the edge of dispersing chromosomes and progressively fuse to form a bilaminar envelope (Fig. 8.13). This results in the formation of chromosome-containing vesicles, i.e. the chromsomes are delimited by two parallel membranes structurally similar to the nuclear envelope (Fig. 8.13b,c). Subsequently, the chromosome-containing vesicles coalesce. This coalescence involves

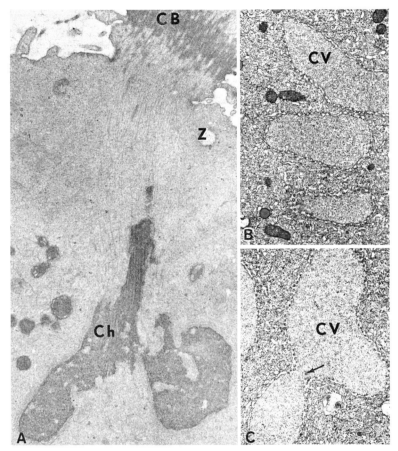

Figure 8.13 Formation of the female pronucleus. (A) chromosomes (Ch) remaining in a rabbit zygote (Z) following completion of the second meiotic division. CB, cytoplasmic bridge connecting the zygote to the second polar body. (B) formation of chromosome-containing vesicles (CV) in a surf clam (*Spisula*) zygote. (C) fusion (arrow) of chromosome-containing vesicles (CV) to form a female pronucleus in a *Spisula* zygote. See Longo (1973).

the fusion of the inner and outer laminae of the chromosome-containing vesicles, thereby forming an irregularly shaped female pronucleus. These events are similar to those observed during the formation of the nucleus in the telophase of mitotic cells.

Subsequent to its formation, the female pronucleus becomes spheroid and may acquire nucleoli, intranuclear annulate lamellae or crystalline structures (Longo, 1973). In many forms, it is difficult to distinguish the female pronucleus from the male and identification is based primarily

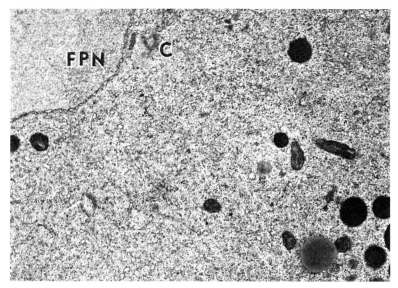

Figure 8.14   Portion of a female pronucleus (FPN) adjacent to a centriole (C) in a mussel (*Mytilus*) zygote. See Longo and Anderson (1970a).

on the proximity of one pronucleus to the site of polar body formation. In some mammals the pronuclei may be identified by size difference or by association with the sperm tail (Austin, 1961). In the sea urchin, *Arbacia*, the male pronucleus is distinguished by the presence of a centriolar fossa, its relatively more electron-opaque chromatin and smaller diameter.

During the formation of the female pronucleus, the aster and portion of the spindle remaining in the zygote regress and may disappear. In the molluscs, *Spisula* and *Mytilus*, the female pronucleus is usually associated with an area relatively devoid of large cytoplasmic components that contains endoplasmic reticulum and some microtubules. Occasionally, a centriole is observed in this region which is believed to be a remnant of the inner aster of the second meiotic apparatus (Fig. 8.14). This structure persists in *Spisula* and *Mytilus* zygotes during pronuclear migration and may be involved in the development of the mitotic spindle and asters.

Cytoplasmic rearrangements during fertilization/meiotic maturation

During fertilization/meiotic maturation, and usually in concert with polar body formation, zygote modifications occur in the form of crenations along the vegetal pole, polar lobes or the redistribution of cytoplasmic components (Raven, 1966; Conrad and Williams, 1974a,b; Dohmen and Van Der Mey, 1977). These morphogenetic events, that are considered to be part of the general phenomenon of cytoplasmic localization/ooplasmic segregation, often result in the redistribution or localization of ooplasmic components (Davidson, 1976). The eggs of many animals do not demonstrate these modifications; the most dramatic examples are seen in eggs that develop polar lobes during meiotic maturation (e.g. the mud snail, *Ilyanassa*).

Many studies consider polar lobe formation only at the time of cleavage. Nevertheless, polar lobes also occur during meiotic maturation. In *Ilyanassa*, the first polar lobe forms when the first meiotic apparatus moves to the animal pole; it recedes upon formation of the first polar body. Just prior to the formation of the second meiotic apparatus, the second polar lobe forms and recedes following the formation of the second polar body.

Cytoplasmic streaming and the redistribution of cellular inclusions in zygotes undergoing meiotic maturation have been described for a number of animals. In general, the movement of cytoplasmic inclusions is believed to be influenced by components in the egg cortex (Raven, 1966; Luchtel, 1976). Aspects involving how these changes are brought about, their correlation with maturation events of the egg, the localization of ooplasmic components and possible cytoskeletal alterations have been reviewed (Jeffrey, 1984).

9

Metabolic alterations at egg activation

Numerous physiological changes occur at fertilization that profoundly affect the activity of the egg, e.g. changes in permeability of small molecules, oxygen uptake, carbohydrate metabolism and synthesis of DNA, RNA and protein. In review of these areas consideration is often given to changes that occur not only at fertilization but throughout embryogenesis (Giudice, 1973; Van Blerkom, 1977; Raff, 1980).

Investigators have demonstrated that the sea urchin egg undergoes permeability changes to different molecules following activation, e.g. its permeability to amino acids and nucleosides increases after fertilization (Giudice, 1973). Active mechanisms for the transport of amino acids and nucleosides are expressed at fertilization and, with respect to gylcine and thymidine, are sodium dependent. The mechanism of transport activation requires an early event of activation (possibly cortical granule exocytosis) and a later event (possibly gradients of Na^+ and K^+) involving increased energy metabolism (Schneider, 1985). Similar permeability changes do not necessarily occur in organisms other than sea urchins and may differ significantly in their mode of activation. In the oyster, *Crassostrea*, fertilized eggs take up uridine at the same rate as unfertilized ova. In the surf clam, *Spisula* there is an increase in amino acid uptake of 10- to 12-fold at the completion of meiotic maturation at about 50 min postinsemination (Bell and Reeder, 1967).

In mice the rate of amino acid uptake is low and relatively constant from the one-cell stage to the blastocyst stage *in vivo*; there are no apparent qualitative or quantitative differences between unfertilized and fertilized eggs (Holmberg and Johnson, 1979). Amino acid uptake in mouse eggs is a carrier-mediated process and that which is accumulated is available for exchange via a carrier-mediated exchange system. Apparently, there is no insertion or activation of amino acid transporting enzymes of similar or novel kinetic activity at fertilization – as appears to be the case with sea urchins. In mouse embryos the uptake of uridine and adenine increases with development but the rates are strikingly different for each nucleoside (Daentl and Epstein, 1971; Epstein and Daentl, 1971; Epstein *et al.*, 1971). In the fertilized egg adenine is taken up about 350-fold more efficiently than uridine and its uptake increases about 20-fold by the blastocyst stage. Uridine uptake during the same period, however, increases 300-fold. This difference in uptake rates of the two nucleosides suggests that they are transported by different systems which are regulated independently.

Oxygen uptake and carbohydrate metabolism

At fertilization the rate of oxygen consumption increases rapidly in sea urchins (Giudice, 1973). In other organisms there is little change at fertilization, e.g. the annelids *Sabellaria* and *Nereis*, or respiration is reduced as in *Cumingia* (bivalve mollusc) and *Chaetopterus* (annelid). In *Paracentrotus lividus* (sea urchin) oocytes the respiration rate is slightly higher than in newly fertilized eggs and much higher than in mature unfertilized eggs (Fig. 10.1). In fish and amphibian eggs there is reportedly no change in respiration at fertilization.

In sea urchin eggs the respiratory burst that occurs during fertilization membrane elevation is associated with the production of peroxide which is the substrate for ovoperoxidase (Foerder, Klebanoff and Shapiro, 1978). Ovoperoxidase joins tyrosyl residues in ditryrosyl linkages which serve to cross-link polypeptide chains of the nascent, soft fertilization membrane resulting in its hardening.

What conditions limit respiration in unfertilized sea urchin eggs and how they are reversed at fertilization has not been fully elucidated. The respiratory change is preceded by a several-fold increase in coenzyme NADPH which is apparently generated by a phosphorylation of NAD (Giudice, 1973). This suggests either an activation of the NAD kinase or that the enzyme and its substrate are compartmentalized and prevented from interacting prior to fertilization. In view of the correlation between the NADPH level and synthetic cellular activities, phosphorylation may be important in initiating and controlling biosynthetic processes of the egg.

The level of glycolytic intermediates and hexose phosphate is low in the eggs of some sea urchins which may limit the oxidative breakdown of carbohydrates and respiration. This, together with the presence of large amounts of gylcogen-like material in the unfertilized egg indicates

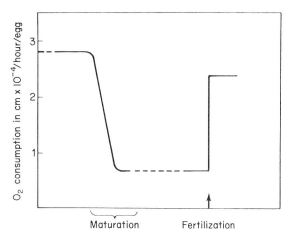

Figure 9.1 Changes in the rate of oxygen consumption during maturation and at the time of fertilization in the sea urchin, *Paracentrotus*. Reproduced with permission from Monroy (1965).

that a block exists in the pathway leading from glycogen to glucose-6-phosphate. This block is apparently reversed at fertilization, as a several-fold increase of glucose-6-phosphate is observed soon after fertilization and shortly before the increased oxygen consumption. These observations suggest that at fertilization there is a mobilization of substrates, possibly due to an activation of the enzyme glycogen-phosphorylase. However, the activity of glycogen-phosphorylase has been found to be at the same level in homogenates of unfertilized and fertilized eggs. The enzyme is released at fertilization from a particulate fraction and may be held in a cellular compartment before fertilization and then released when required.

Intermediates of the tricarboxylic acid cycle are rapidly oxidized by homogenates of sea urchin (*Arbacia*) eggs and isolated mitochondria are capable of oxidative phosphorylation (Giudice, 1973). Pentose cycle activity is enhanced at fertilization and tends to predominate through fertilization and early cleavage.

Metabolic changes in mollusc eggs at fertilization appear to be quite variable depending upon the species examined (Raven, 1972). Glycolysis is regulated at steps of phosphorylase, phosphofructokinase and pyruvate kinase in fertilized *Crassostrea* eggs. Phosphorylase is activated first, followed by pyruvate kinase and phosphofructokinase. Because this pattern is similar to that found in sea urchin eggs it is possible that these rate-limiting steps are regulated in the same manner, and carbohydrate utilization is enhanced at fertilization by an activa-

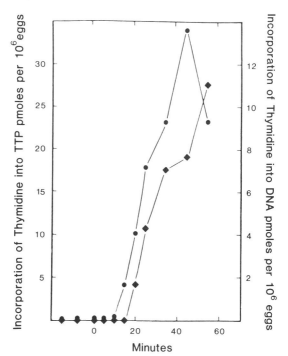

Figure 9.2 Incorporation of ^{3}H-thymidine into thymidine triphosphate (-●-●-) and DNA (-♦-♦-) of sea urchin (*Arbacia*) zygotes. The sperm suspension was added at 0 min. Reproduced with permission from Longo and Plunkett (1973).

tion of glycolysis (Yasumasu, Tazawa and Fujiwara, 1975). Activation of NAD kinase, as inferred from the elevation of total NADP levels in surf clam (*Spisula*) zygotes, occurs immediately following insemination and is regulated by calcium and calmodulin *in vitro* (Epel *et al.*, 1981).

Although respiration and energy utilization increase at fertilization in the sea urchin, *Strongylocentrotus*, there is reportedly no change in the levels of ATP, ADP, and AMP during the period of maximum respiratory activity (Epel, 1969). These results do not support the hypothesis that respiration in eggs is limited by ADP. ATPase activity reportedly increases after fertilization in sea urchins and is calcium activated. A magnesium activated ATPase whose activity increases slightly following fertilization has been described (Monroy, 1957).

DNA synthesis

One of the most dramatic consequences of fertilization is stimulation of the egg to undergo DNA replication and cell division (Fig. 9.2). Nuclear transplantation experiments showed that brain cell nuclei injected into mature amphibian eggs undergo DNA synthesis and this synthesis is replicative rather than repair (Gurdon, Birnstiel and Spright, 1969; Laskey and Gurdon, 1973). In similar experiments with immature oocytes, injected nuclei do not undero DNA synthesis. These observations correlate with the overall increase in DNA polymerase activity during *Xenopus* maturation.

Critical events in the initiation of DNA synthesis at fertilization are unknown. The enzymatic machinery required for synthesis is present and the deoxynucleotide pool, although small, does not appear to be limiting (De Petrocellis and Rossi, 1976). All four deoxyribonucleotide kinases have been directly assayed in the sea urchin, *Strongylocentrotus* (Fansler and Loeb, 1969). Their activity was not found to vary significantly in unfertilized and fertilized eggs. The enzymatic activity of DNA polymerase from homogenates of *Strongylocentrotus* does not undergo variations with phases of the cell cycle. The enzymes (kinase and polymerase) appear to be present in both the nucleus and cytoplasm of the unfertilized egg. In *Strongylocentrotus* the percent DNA polymerase activity associated with the nuclear fraction increases progressively while activity in the cytoplasm declines. By the late blastula stage most of the DNA polymerase activity is associated with the nucleus. This change in DNA polymerase activity has been interpreted to represent a transfer of the enzyme from the cytoplasm to the nucleus as development proceeds (Fansler and Loeb, 1972; Loeb *et al.*, 1969; Loeb and Fansler, 1970).

DNA polymerase activity is not detected in sea urchin spermatozoa, suggesting that the ability of male pronuclei to undergo DNA synthesis is due to the association of DNA polymerase of maternal origin with the paternally derived DNA during pronuclear development. In mammals, RNA and DNA polymerase activities have been shown to be associated with the mitochondria of bull sperm (Hecht, 1974; Hecht and Williams, 1979).

DNA synthesis in sea urchin eggs has been experimentally inhibited by a number of different agents with delays or complete cessation of cleavage (Giudice, 1973). DNA synthesis is inhibited by substances that interfere primarily with protein synthesis, such as puromycin and cycloheximide, suggesting that some proteins associated with DNA synthesis are synthesized *de novo*. Although these drugs block cell division they have little effect on pronuclear development and association.

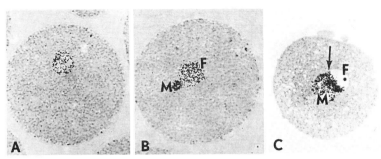

Figure 9.3 Autoradiographs of sea urchin and rabbit zygotes incubated in ³H-thymidine. (A) Autoradiographic grains over the zygote nucleus of a sea urchin (*Arbacia*) zygote. In this species DNA synthesis normally follows pronuclear fusion. (B) *Arbacia* zygote treated with colchicine to inhibit pronuclear fusion. Both the male (M) and the female (F) pronuclei have synthesized DNA. (C) Closely apposed male (M) and female (F) pronuclei of a rabbit zygote that have synthesized DNA. Autoradiographic grains in the female pronucleus are located at the pole (arrow) closest to the male pronucleus. See Longo and Plunkett (1973) and Longo (1976c).

Although the unfertilized sea urchin egg can take up thymidine, it is unable to phosphorylate the nucleoside to thymidine triphosphate (TTP; Fig. 9.2). Phosphorylation of thymidine to TTP begins about 10 min postinsemination in *Arbacia* zygotes (Longo and Plunkett, 1973). Since homogenates of unfertilized eggs are unable to carry out thymidine phosphorylation, there may be a compartmentalization of phosphorylation enzymes, thereby preventing their interaction with appropriate substrates. In *Arbacia*, DNA synthesis normally follows pronuclear fusion, and occurs in the zygote nucleus about 16 min postinsemination at 20°C (Fig. 9.3). In sand dollar, mouse, and rabbit zygotes DNA synthesis occurs in both pronuclei during their migration (Simmel and Karnofsky, 1961; Oprescu and Thibault, 1965; Luthardt and Donahue, 1973). There is a distinctive distribution of silver grains over the female pronucleus in autoradiographs of rabbit zygotes (Fig. 9.3). The grains are distributed in a polarized fashion and located along that region of the female pronucleus proximal to the male pronucleus. This distribution corresponds to the localization of DNA within the female pronucleus as demonstrated by the Feulgen technique.

There is little doubt that DNA synthesis is controlled by cytoplasmic factors in the egg which are activated at fertilization. When DNA synthesis begins before pronuclear fusion, it occurs simultaneously in both pronuclei. If eggs are made polyspermic or injected with accessory sperm, DNA synthesis begins at the same time in all pronuclei (Graham

1966; Longo and Plunkett, 1973). When ascidian (*Ascidia*) eggs are cut into animal and vegetal halves and then fertilized, DNA synthesis begins at the same time in both halves (Ortolani *et al.*, 1975). These results suggest that fertilization triggers a chain of reactions that rapidly propagate throughout the cytoplasm and which result in the removal of the block to DNA synthesis.

The observation that DNA synthesis can be initiated and maintained by exposure of unfertilized sea urchin eggs to ammonia gave rise to the idea that DNA synthesis is activated by an increase in intracellular pH (Mazia and Ruby, 1974). Although eggs activated in ammonia and calcium-free sea water undergo an increase in internal pH, they do not initiate DNA synthesis. This suggests that ammonia activation is more than just an elevation of internal pH. It appears that both calcium and pH elevation are required for the initiation of DNA synthesis in sea urchin eggs (Whitaker and Steinhardt, 1981).

RNA and protein synthesis

The existence of informational macromolecules in the egg cytoplasm is suggested by studies of developing marine embryos that demonstrate the presence of morphogenetic factors and that various regions of the uncleaved ovum are not equivalent in their developmental potential (Davidson, 1976). Harvey (1956) produced egg fragments by centrifugation and showed that parthenogenetic merogones (enuclate egg fragments) were capable of developmental changes. Studies of hybrid embryos also showed that in some forms the genome of the early embryo is silent. Later studies using biochemical rather than morphological criteria, demonstrated that the eggs of sea urchins and other organisms possess a store of messenger RNA that is translated subsequent to fertilization.

Incorporation of amino acid into protein by non-nucleate fragments of sea urchin eggs can be stimulated by activation to the same extent as by fertilization which suggests the presence of mRNA in the egg cytoplasm (Brachet *et al.*, 1963; Denny and Tyler, 1964). Inhibition of transcription in sea urchin embryos treated with actinomycin D does not prevent DNA or protein synthesis and cleavage (Fig. 9.4) and treated embryos are capable of developing as far as the hatching blastula stage (Gross and Cousineau, 1964). An implicit assumption of these studies is that actinomycin D affects only transcription (Sargent and Raff, 1976). Actinomycin D experiments have also been carried out with embryos of snails and tunicates with results comparable to those in

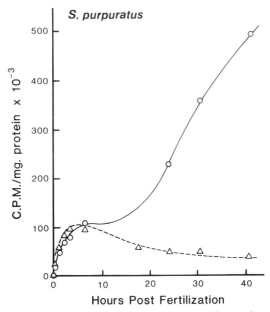

Figure 9.4   Effect of actinomycin D on protein synthesis of sea urchin (*Strongylocentrotus*) embryos determined by ^{14}C-valine incorporation. ○, controls. △, embryos exposed continuously to 20 μg/ml actinomycin D from before fertilization. The second rise in controls begins after the hatching blastula stage and does not occur in specimens treated with actinomycin D. Rates of actinomycin D treated specimens are slightly higher for the first 2 to 3 h than the controls. Reproduced with permission from Gross (1964).

sea urchins. Mammalian embryos continue protein synthesis in the presence of actinomycin D with an inhibition of RNA synthesis (Manes, 1973; Golbus, Calarco and Epstein, 1973). Similar results have been obtained in mammalian embryos with α-amanitin that inhibits RNA synthesis but neither blocks cleavage nor inhibits protein synthesis.

The sea urchin embryo can reach the blastula stage in the absence of RNA synthesis, however, this does not imply that RNA is not normally synthesized during this early period. In addition to an incorporation of precursors into the pCpCpA sequence of tRNA, the synthesis of heterogeneous RNA does, in fact, take place in the period from fertilization to hatching in sea urchins (Gross, Kraemer and Malkin, 1965). Moreover, there is evidence that the paternally-derived genome may be active in RNA synthesis during fertilization (Longo and Kunkle, 1977).

The synthesis of RNA upon fertilization has been investigated in molluscs (Collier, 1976; Kidder, 1976; McLean, 1976). Fertilized oyster (*Crassostrea*) eggs take up uridine at the same rate as unfertilized eggs, and the rate at which uridine is incorporated into high molecular weight RNA is not altered at insemination (McLean and Whiteley, 1974). Hence, development of *Crassostrea* during cleavage is apparently controlled by information stored in the unfertilized egg. Although RNA synthesis is not markedly increased at fertilization in the snail, *Lymnaea*, all major forms of RNA are synthesized from oogenesis throughout fertilization. In contrast to these results, eggs and early embryos of the gastropod, *Acmaea*, reportedly do not incorporate uridine until late cleavage (Karp, 1973).

Incorporation of the nucleosides uridine, cytosine and adenine into pronuclei of mouse zygotes was demonstrated by Mintz (1964), although later studies using ^{3}H-uridine failed to detect RNA polymerase activity *in vitro*. However, transcriptional activity at the pronuclear stage in mouse zygotes has been demonstrated using ^{3}H-adenine, possibly due to the more efficient uptake of adenine vs. uridine (Clegg and Pikó, 1982). The absolute rate of RNA synthesis increases about twofold as mouse embryos progress from the one- to the eight-cell stage of development; much of this newly synthesized RNA is attributable to enhanced synthesis of ribosomal RNA and the production of ribosomes. Correlated with the onset of detectable rRNA synthesis is the modification in nucleolar structure. With the appearance of granule elements and the progressive reticulation of the nucleolar matrix, ribosomes and polysomes become more abundant within the cytoplasm (Van Blerkom and Motta, 1979).

A number of specific proteins including tubulins and the major histones, have been identified as products of translation of stored mRNAs in sea urchins (Raff *et al.*, 1971; Ruderman and Gross, 1974). Investigations of histone mRNA have shown that it is localized within the female pronucleus (Showman *et al.*, 1982). The reason for this compartmentalization of histone mRNA has not been determined.

Unfractionated RNA from unfertilized sea urchin eggs has been translated in cell-free systems indicating that 3–4% of the egg RNA is mRNA (Raff, 1980). RNA extracted from 20 to 40 S particles of sea urchin eggs directed the synthesis of histones *in vitro* (Gross *et al.*, 1973). Similar results have also been obtained from surf clam (*Spisula*) embryos. Sea urchin eggs contain poly(A)$^+$ mRNA, poly(A)$^-$ histone mRNA and poly(A)$^-$nonhistone mRNA. Amphibian oocytes contain poly(A)$^+$ mRNA, poly(A)$^-$ and poly(A)$^+$ histone mRNA. Observations of Ruderman and Pardue (1977) suggest that sea urchin embryos, in

Table 9.1 Comparison of changes in the rates of protein synthesis and respiration at fertilization for selected species of vertebrates and invertebrates. See Houk and Epel (1974).

Species	Maturation stage when fertilized	Increase in rates	
		Protein synthesis	Respiration
Urechis caupo (echiuroid)	Intact germinal vesicle (GV)	At fertilization, 2×	At fertilization, 1.2×
Spisula solidissima (mollusc)	Intact GV	At fertilization, 3–4×	Not at fertilization
Sea urchins, many species	Pronucleate egg	At fertilization, 6–30×	At fertilization, 6–10×
Rana pipiens (amphibian)	Second metaphase	At GV breakdown, 10×	Not at fertilization
Asterias forbesii (asteroid)	After GV breakdown	After GV breakdown	At completion of meiosis
Patiria miniata (asteroid)	After GV breakdown	Before GV breakdown, 5×	At completion of meiosis

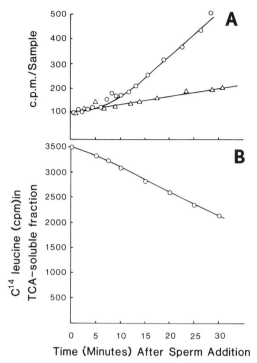

Figure 9.5 (A) Incorporation of ^{14}C-leucine in fertilized sea urchin (*Lytechinus*) eggs. Sperm was added at time zero. Unfertilized eggs were preloaded with ^{14}C-leucine; one part of the sample was fertilized (○), the other was left unfertilized (△). Samples collected at timed intervals were analyzed for incorporation of the label in hot trichloroacetic acid-insoluble material. (B) Loss of ^{14}C-leucine from the free amino acid pool in fertilized eggs. Reproduced with permission from Epel (1967).

contrast to eggs, contain relatively few prevalent nonhistone poly(A)$^-$ mRNAs and that newly synthesized nonhistone poly(A)$^-$ mRNAs contribute less (at least quantitatively) to the embryonic program than do the poly(A)$^+$ mRNA and histone mRNA components. Sequence complex analysis has been carried out for eggs of sea urchins, amphibians (*Xenopus*) and the echiuroid worm, *Urechis* (Davidson, 1976). RNAs corresponding to the single copy portion of the genome are present in oocytes of these organisms, equivalent to about 25 000 different mRNAs of 1500 nucleotides in length.

Studies of changes in the rate of protein synthesis for eggs inseminated at different stages of meiosis are summarized in Table 9.1. In sea urchin eggs a large spectrum of proteins is synthesized at a comparatively low rate (Brandhorst, 1976). However, fertilization is followed by a several-fold increase in the rate of protein synthesis which indicates the release of the block restricting mRNA translation in the unfertilized egg (Fig. 9.5.). However, this increase in rate is not accompanied by a significant change in the pattern or number of polypeptides synthesized. The protein synthesis increase is independent of mRNA synthesis and is accompanied by an increase in polysomes at the expense of single 80 S ribosomes, as stored maternal mRNA is recruited in a linear manner following fertilization. Possible explanations for the dramatic post-fertilization rise in protein synthesis in sea urchin eggs have been reviewed by Raff (1980) and include: (a) a faulty or incomplete translation apparatus, (b) mRNA processing or modification, and (c) mRNA unmasking. Unmasking appears to be the primary mechanism affecting the change in protein synthesis at fertilization.

The low rate of protein synthesis in unfertilized sea urchin eggs does not appear to be due to a lack of ribosomes or energy sources. Relatively little is known about the involvement of various factors controlling initiation, elongation, and termination of polypeptide chains during protein synthesis in fertilized eggs. A translation inhibitor has been isolated from egg ribosomes with salt solutions and suppresses the translation of poly(U) templates *in vitro*, suggesting that this factor may act to inhibit translation in unfertilized eggs. However, Hille (1974) showed that the inhibitor could be isolated from active embryo ribosomes. Moreover, egg ribosomes, that presumably possess the inhibitor, can translate globin mRNA as well as embryo ribosomes (Clegg and Denny, 1974).

The modification or processing of mRNA at fertilization has not been extensively studied. The inverted 'cap' of 7-methylguanosine at the 5′ end of eukaryotic mRNAs has been investigated. In sea urchins a significant portion of egg mRNAs possess this structure (Raff, 1980). Capping of mRNAs could provide a store of non-translatable message since mRNA becomes translatable after methylation. A second type of modification found on the 3′ end of many mRNAs is poly(A) segments which in the case of sea urchin eggs, increase twofold within 2 h of fertilization. However, blockage of 3′polyadenylation of mRNA with codycepin has no apparent effect on the rise of protein synthesis or polysome formation. Histone mRNA, which lacks poly(A) segments in sea urchin eggs, participates in the postfertilization rise in protein synthesis. Thus, polyadenylation may not represent either the factor which controls the interaction of ribosomes with mRNA or the condition that makes the maternal mRNA translatable. Experiments comparing the rate of translation of poly(A)$^+$ and poly(A)$^-$ globin mRNA injected into amphibian (*Xenopus*) oocytes demonstrate that the initial rate of globin synthesis is similar in both cases (Huez *et al.*, 1974). With longer incubation the rate of globin synthesis directed by poly(A)$^-$mRNA, is considerably lower than that directed by poly(A)$^+$mRNA. These observations suggest that poly(A)$^+$ sequences may increase the stability of mRNA.

A widely held hypothesis is that the protein synthesis machinery in eggs is complete and that the mRNAs are potentially translatable but they are sequestered or masked and, therefore, unavailable for translation (Raff, 1980). Fertilization/artificial activation triggers unmasking of mRNAs. Consistent with this hypothesis are studies demonstrating that the mRNAs found in presumptive, cytoplasmic, messenger ribonuclear particles (mRNPs) exhibit S values greater than those of the same mRNAs after purification. For example, histone mRNPs typically have S values of 9–12, but sediment at 20–60 S in egg homogenates. In addition, purified poly(A)$^+$ mRNAs, which have model S values of 20–30, are found to sediment at 30–70 S when detected in egg homogenates.

The existence of mRNPs in the cytoplasm of unfertilized and fertilized eggs, although consistent with the hypothesis of masked mRNA, is not sufficient to substantiate the existence and developmental role of masked mRNA. Messenger RNPs isolated from polysomes or cytoplasm of various systems are as effective as the isolated mRNAs in stimulating heterologous *in vitro* protein synthesis (Raff, 1980). Moreover, demonstration of the validity of the masking hypothesis requires that egg mRNPs, isolated in their native state, be nontranslatable by an *in vitro* protein synthesizing system until they have been modified. This modification allows the contained mRNA to be translated.

The translatability of poly(A)$^+$ mRNPs from sea urchin eggs has been examined (Jenkins *et al.*, 1978; Young and Raff, 1979). Cytoplasmic poly(A)$^+$ mRNPs fail to stimulate translation in a wheat germ system, whereas mRNA extracted from these non-translatable particles was shown to be as template active as mRNA extracted from whole eggs. The lack of translation of egg mRNPs may not be due to the presence of an inhibitor, since addition of mRNPs to deproteinized mRNA has no effect on the translation efficiency of the latter.

As a working hypothesis, Humphreys (1969, 1971) suggested that the rate of protein synthesis may be increased at fertilization by a heightened efficiency of translation and/or the translation of additional mRNA molecules. The efficiency of translation, defined as the number of protein molecules produced per mRNA molecule per unit time, is similar in sea urchin eggs and embryos. Protein synthesis is believed to be accelerated at fertilization by the translation of additional mRNA molecules. Measurements of mRNA entering polysomes in fertilized sea urchin eggs, revealed that ribosomes in polysomes increase about 30-fold from 0.75% to 20% following fertilization, so substantiating Humphreys' speculation. These results suggest that the translational control mechanism in the egg acts directly at the level of the mRNA molecule. Possible control mechanisms resulting in the activation of mRNA include: (a) alterations in a molecule associated with the mRNA, (b) structural changes in the mRNA itself, and (c) release of mRNA from a sequestered cellular compartment.

By studying the translational activity of sea urchin ribosomes in a fractionated reticulocyte cell-free system Danilchik and Hille (1981) have determined that unfertilized egg ribosomes differ from activated egg ribosomes. In comparison to ribosomes from embryos, those from eggs are less active in polymerizing amino acids and slowly increase this activity over the course of incubation. Furthermore, differences between egg and blastula ribosome activity may be involved in postactivation of protein synthesis. Consistent with Danilchik and Hille's observations are those showing a quantitative loss of high molecular weight protein from *Strongylocentrotus* egg ribosomes following fertilization (Unsworth and Kaulenas, 1975). Investigations demonstrating that the ribosome transit time (the time necessary to traverse a mRNA during translation) decreases by more than one half in fertilized sea urchin eggs, offers another potential challenge to the view that the protein synthesis increase at fertilization results solely from recruitment of stored messenger RNA (Brandis and Raff, 1978; Hille and Albers, 1979). However, the magnitude of such changes can only account for a small proportion of the increased synthetic activity. Hence, it appears that the rate of protein synthesis in eggs and zygotes is primarily controlled by availability of mRNA. A pool of masked mRNA may be maintained in the egg and upon fertilization unmasking begins and translational efficiency rises. The result is the recruitment of mRNA into polysomes and a concomitant increase in protein synthesis. In sea urchins this process may involve modification or substitution of proteins in the mRNPs.

In the mud snail, *Ilyanassa*, protein synthesis occurs in the unfertilized egg; approximately 15 min post-fertilization, amino acid uptake and incorporation increase. When the percentage of incorporation is calculated, protein synthetic activity of the zygote is about 2.5 times greater than the unfertilized ovum (Mirkes, 1970). Rapid and dramatic changes in the pattern of protein synthesis occur in fertilized surf clam (*Spisula*) eggs (Rosenthal, Hunt and Ruderman, 1980). There is a synthesis reduction of prominent oocyte-specific proteins and a synthesis increase for at least three proteins whose labeling dominate the pattern of protein synthesis in early embryos. Alterations in the pattern of protein synthesis at fertilization in *Spisula* are due to stage-specific utilization of different subsets of mRNA from a common maternal pool. Discrimination of mRNAs may be achieved by a selective repression of availability by a phenol-soluble component of the egg.

A small number of polysomes have been shown to be active in protein synthesis in *Spisula* eggs (Firtel and Monroy, 1970). After fertilization, there is a progressive increase in the number of ribosomes that become associated with polysomes, and by 30 min postinsemination the specific activity of polysomes is 2.5 times greater than in unfertilized ova. By the time the pronuclei are formed there are four to five times more polysomes than were present in the egg prior to maturation. These results suggest that the increase in the number of polysomes at fertilization is due, at least in part, to the activation of stored, maternally derived mRNA. Eggs of the coot clam, *Mulinia*, also undergo an increase in polysomes by 45 min postinsemination (Kidder, 1976). Furthermore, there does not seem to be a shift in polysome size and distribution at fertilization. Protein synthesis in the molluscs, *Spisula* and *Mulinia*, therefore, is similar to that observed in sea urchins, suggesting that gene transcription patterns between early mosaic and regulative embryos are not significantly different.

Studies with amphibian (*Rana pipiens*) oocytes have demonstrated that an increased rate of protein synthesis (70%) follows the onset of maturation; fertilization results in a further increase (50%) and by the blastula stage the rate has increased an additional twofold (Shih *et al.*, 1978). In *Xenopus* eggs the polysome content increases an additional twofold shortly after fertilization.

Accompanying the final meiotic division of the mammalian oocyte are qualitative and quantitative changes in the pattern of protein synthesis (Van Blerkom, 1977; Sherman, 1979; Wassarman *et al.*, 1981). The majority of polypeptides synthesized by mature unfertilized mouse eggs are also made in the zygote and some show a change in the relative rate of synthesis following insemination. In addition, during the immediate post-fertilization period of both rabbit and mouse embryos, major changes in the pattern of protein synthesis take place. Although differential mRNA recruitment may account for some of these changes many result from post-translational modifications such as phosphorylation and glycosylation (Van Blerkom, 1981; Howlett, 1986.)

Certain proteins, e.g. tubulin and ribosomal, are synthesized in large amounts by growing mouse oocytes and continue to be synthesized at similar or even greater rates during early embryogenesis (Wassarman *et al.*, 1981). In rabbit eggs there is a translation of a population of stage-specific polypeptides that is autonomous of fertilization and appears to follow a timed, translational schedule initiated with germinal vesicle breakdown (Van Blerkom, 1979). While tubulin is synthesized from mRNA stored in unfertilized sea urchin eggs, it is unclear to what extent maternally derived tubulin mRNA directs tubulin synthesis during mouse embryogenesis (Wassarman *et al.*, 1981). Like tubulin, ribosomal protein synthesis represents a major portion of the total proteins synthesized in mouse eggs and, in combination with rRNA synthesis, this suggests that the unfertilized mouse ovum, like eggs of non-mammalian species, contain a store of ribosomes for use during embryogenesis.

Direct comparisons between intracellular pH and protein synthesis in eggs and early sea urchin embryos indicate that pH regulates protein synthesis in a reversible manner (Epel *et al.*, 1974; Grainger *et al.*, 1979). Calcium release in the absence of an intracellular pH increase does not stimulate protein synthesis, while a pH increase in the absence of calcium release yields a partial stimulation. Only in the presence of intracellular pH increase and calcium release is the rate of protein synthesis in experimentally treated ova comparable to that of fertilized eggs (Winkler *et al.*, 1980). The mechanism by which these ionic signals activate protein synthesis in unknown. A cell-free system derived from the sea urchin, *Lytechinus pictus*, exhibits the ionic controls found *in vivo* and is capable of initiation, elongation and termination of the normal spectrum of proteins synthesized *in vivo* at physiological ion concentrations. However, this system has a relatively low and variable activity which suggests that it is incomplete.

10

Development of the male pronucleus

Although specific details involving the development of the male pronucleus vary from one organism to another, there are three basic features of this process: (a) breakdown of the sperm nuclear envelope, (b) dispersion of the condensed sperm chromatin, and (c) development of a nuclear envelope, the pronuclear envelope. These events are accompanied by a dramatic transformation in the shape, volume, chromatin conformation, nucleoprotein content and activity of the incorporated sperm nucleus (Longo, 1973, 1981b; Longo and Kunkle, 1978).

Breakdown of the sperm nuclear envelope

The sperm chromatin of some animals is not delimited by a nuclear envelope (Baccetti and Afzelius, 1976). Following gamete fusion in these instances the paternally-derived chromatin is placed in direct association with the egg cytoplasm, without an intervening membranous boundary. However, the sperm of most organisms do contain a nuclear envelope. Immediately following incorporation the inner and outer laminae of the sperm nuclear envelope fuse at multiple sites, thereby forming vesicles that initially outline the condensed sperm chromatin but are then scattered throughout the surrounding cytoplasm (Fig. 10.1). The vesicles lack distinguishing features and are soon lost amongst other membranous elements. As a result of the breakdown of the sperm nuclear envelope, the condensed sperm chromatin is directly exposed to the zygote cytoplasm (Longo, 1973).

A number of observations indicate that breakdown of the sperm nuclear envelope at fertilization is a highly regulated event. For example, in the sea urchin, *Arbacia*, vesiculation of the sperm nuclear envelope is not complete (Longo, 1973). Segments of the sperm nuclear

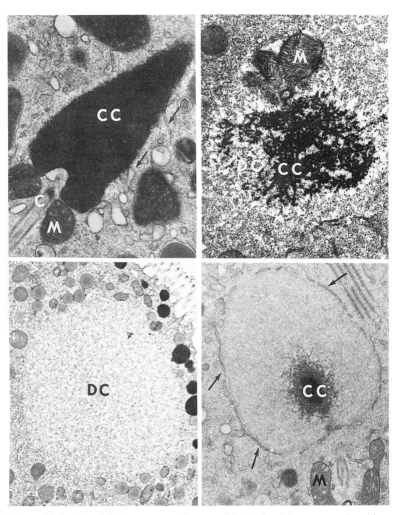

Figure 10.1 (*top left*) Incorporated sea urchin (*Arbacia*) sperm nucleus that has undergone sperm nuclear envelope breakdown. Vesicles (arrows) adjacent to the condensed sperm chromatin (CC) may be derived from the vesiculation of the sperm nuclear envelope. M, sperm mitochondrion; C, centriole.

Figure 10.2 (*top right*) Incorporated surf clam (*Spisula*) sperm nucleus undergoing chromatin dispersion. DC and CC, dispersed and condensed chromatin; M, sperm mitochrondia. See Longo and Anderson (1970b)

Figure 10.3 (*bottom left*) Dispersed sperm chromatin (DC) of a surf clam (*Spisula*) zygote prior to the formation of the male pronuclear envelope.

Figure 10.4 (*bottom right*) Development of the male pronuclear envelope in a sea urchin (*Arbacia*) zygote. The vesicles (arrows) surrounding the dispersed sperm chromatin fuse to form a continuous lamina. CC, condensed chromatin; M, sperm mitochondrion.

envelope formerly associated with the acrosome and centrioles are left intact; subsequently they become incorporated into the nuclear envelope of the male pronucleus. Retention of specific regions of the sperm nuclear envelope has also been observed in mammalian zygotes (Yanagimachi and Noda, 1970a). In *Arbacia*, breakdown of the sperm nuclear envelope sometimes occurs in the immediate vicinity of the female pronucleus. In this instance the nuclear envelope of the female pronucleus fails to undergo similar changes, indicating the specificity of this process.

Although factors regulating the disappearance of the sperm nuclear envelope have not been determined, it is possible that processes similar to nuclear envelope disruption in mitotic cells are involved. Mature amphibian ova contain a cytoplasmic factor that can induce germinal vesicle breakdown when injected into immature oocytes (Masui and Clarke, 1979). The activity of this factor is quickly lost after fertilization but later reappears and cycles with cleavage (Wasserman and Smith, 1978). Its appearance coincides with the onset of mitosis in cycling cells, suggesting that the factor may not be restricted to maturing oocytes but may play a more general role in regulating nuclear envelope breakdown.

Chromatin dispersion

Transformation of the condensed sperm chromatin into the dispersed form of the male pronucleus is a dramatic alteration that has been examined in a variety of organisms (Longo, 1973). Morphological changes usually occur first along the periphery of the condensed, incorporated sperm nucleus; dense staining chromatin grades into a more dispersed and lightly staining mass (Fig. 10.2). As this process continues, the peripheral dispersion zone increases in volume, while the central dense portion gradually decreases until it disappears. One interpretation of such a pattern of morphogenesis is that the agent(s) responsible for dispersion initiates the process at the periphery; once the outer chromatin is dispersed the more central chromatin is free to disperse also. Eventually, all the sperm chromatin becomes a morphologically homogeneous mass of dispersed chromatin (Fig. 10.3). In some cases, chromatin dispersion appears to occur uniformly and simultaneously throughout the sperm nucleus, without the formation of particular regions with different densities and conformations. As a result of dispersion, the paternal chromatin undergoes a significant increase in volume.

The pattern of chromatin dispersion appears to partially govern the initial structure of the male pronucleus. For example, in the sea urchin, *Arbacia*, the retention of the nuclear envelope along the apical and basal regions of the sperm nucleus and the lateral dispersion of sperm chromatin yields a heart-shaped mass which, following the development of the pronuclear envelope, becomes a spheroid male pronucleus. In mammals the dispersed chromatin profile is ellipsoidal and is reminiscent of the original shape of the sperm nucleus.

It has been speculated that sperm chromatin dispersion in mammals is the opposite of nuclear condensation during spermiogenesis (Szollosi and Ris, 1961; Bedford, 1970). If this were the case, it would be suggestive of a reversible process regulating the paternal chromatin at spermiogenesis and at fertilization. However, morphological and biochemical analyses of paternal chromatin during pronuclear development and spermiogenesis indicate that this is not the case in invertebrates.

It is not unreasonable to suggest that dispersion of the condensed sperm chromatin is a morphological manifestation of changes in nucleoprotein content; in fact, chemical alterations are coincident with sperm chromatin dispersion. Formidable problems are inherent in studies of chemical changes in the paternal chromatin at fertilization due to technical difficulties, e.g. isolation of pronuclei is arduous because of an extremely high cytoplasm/nucleus ratio characteristic of the zygote. Despite these difficulties, a number of studies have been performed that yielded interesting results.

In many organisms, during the differentiation of the spermatogonium into a spermatozoon, histones characteristic of somatic cells are replaced by a distinct group of basic nuclear proteins, often unique to the mature spermatozoon and more basic than those found in somatic cells. It has been suggested that the complexing of sperm DNA to these basic proteins permits condensation of the chromatin and repression of the DNA (Bloch, 1969).

Cytochemical examinations of fertilized eggs demonstrate that the paternal chromatin stains differently subsequent to the metamorphosis of the sperm nucleus into a male pronucleus which suggests that the DNA brought into the egg with the spermatozoon acquires different basic nucleoproteins, presumably during the dispersion of the sperm chromatin (Bloch and Hew, 1960; Das, Micou-Eastwood and Alfert , 1975). Support for the idea that the sperm basic nuclear proteins are removed from paternal DNA comes from autoradiographic analyses of incorporated sperm nuclei labeled with amino acids (Ecklund and Levine, 1975). During the differentiation of the sperm nucleus into a male pronucleus there is a reduction in autoradiographic grains associated with the dispersing chromatin, indicating that basic proteins unique to the sperm nucleus are not simply 'masked' but are dissociated from the DNA. Similarly, antibodies to mouse sperm basic proteins fail to detect antigenic sites in fertilized eggs (Rodman *et al.*, 1981).

Biochemical studies of transforming, paternal chromatin at fertilization indicate that the basic proteins of the sperm nucleus are lost and the paternal DNA associates with basic proteins similar to those found within the female pronucleus (Carroll and Ozaki, 1979; Poccia, Salik and Krystal, 1981). This transformation occurs in the absence of protein synthesis, suggesting a pool of maternal basic proteins is available for this conversion. As in the case of basic nucleoproteins, the nonbasic nucleoproteins of the spermatozoon are apparently replaced by ones similar to those found within the female pronucleus (Kunkle, Longo and Magun, 1978).

Sperm nuclear decondensation activity of egg cytoplasm has been examined in sea urchins and amphibians by mixing sperm nuclei and egg homogenates *in vitro* (Kunkle, Magun and Longo, 1978; Eng and Metz, 1980; Lohka and Masui, 1983a,b, 1984a). Homogenates and cytosol preparations of fertilized and unfertilized eggs induce changes in sperm nuclei similar to those observed *in vivo*. In the sea urchin, *Lytechinus*, this activity appears to be associated with a protein(s) of approximately 10^5 molecular weight. In amphibians there is evidence that sperm nuclear decondensation *in vitro* requires calcium and cellular components other than soluble factors. Interestingly, the addition of proteolytic inhibitors (soybean trypsin inhibitor, ovomucoid and phenylmethylsulfonylfluoride) to egg homogenates or cytosol preparations do not inhibit sperm decondensation *in vitro*.

Investigations with mammalian sperm *in vitro*, demonstrate that cleavage of disulfide bonds is required for decondensation, and thereby permits the disruption of nucleoproteins bound to DNA (Calvin and Bedford, 1971). A combination of a reducing agent, such as dithiothreitol, and urea or trypsin, has been employed to induce dispersion. The extreme conditions affected by such combinations may mimic but undoubtedly do not reflect the biology of *in vivo* decondensation. A high concentration of thiol groups and glutathione reductase in ova support speculations on their involvement in sperm chromatin dispersion *in vivo* (Wiesel and Schultz, 1981). Phosphorylation is involved in the modification of basic nucleoproteins during spermatogenesis and, in combination with other enzymatic modifications of the nucleoproteins, may be instrumental in dispersion of the condensed sperm chromatin. For example, high levels of phosphorylation in the mouse egg at fertilization and the phosphorylation of sperm protamines by egg extracts suggest that this process could lead to a charge modification and a destabilization of protamine-DNA interactions of the sperm chromatin during male pronuclear development (Young and Sweeney, 1978; Wiesel and Schultz, 1981). In fertilized sea urchin eggs, sperm-specific histones H1 and H2 are phosphorylated. Later, in parallel with chromatin dispersion, nearly all phosphorylated sperm H1 is lost from the pronuclear chromatin, with the concurrent assimilation of egg derived phosphorylated H1 (Green and Poccia, 1985).

Sulfhydryl-induced proteolytic activity has also been proposed to be involved in mammalian sperm nuclear decondensation *in vivo* (Zirkin and Chang, 1977; Zirkin, Chang and Heaps, 1980). An acrosin-like protease, associated with isolated rabbit sperm nuclei, causes nucleoprotein degradation and decondensation *in vitro*. However, the normal role of this proteolytic activity *in vivo* has been questioned, since the decondensing activity intrinsic to isolated sperm nuclei may be of acrosomal origin and may become bound to the chromatin during sperm isolation (Young, 1979). In connection with the possible involvement of proteolytic agents in sperm chromatin dispersion, it is noteworthy that proteolytic activity is associated with the nuclei of somatic cells and processing of histone.

Amphibian and sea urchin eggs have been shown to contain DNA binding proteins and pronuclei of these ova can concentrate cytoplasmic nonhistone proteins (Barry and Merriam, 1972; Kunkle, Magun and Longo, 1978). Interestingly, Barry and Merriam (1972) showed that chick erythrocyte nuclei swell when suspended in cytoplasm from *Xenopus* eggs but are not altered when mixed with cytoplasm from immature oocytes. A significant accumulation of label in male and female pronuclei of rabbit oocytes was found when eggs were incubated *in vitro* with ^{3}H-lysine and subsequently inseminated, providing additional evidence that proteins from the egg cytoplasm are incorporated into both pronuclei at fertilization (Motlik *et al.*, 1980).

Cytoplasmic proteins are recruited into nuclei before swelling and DNA synthesis occur, suggesting a causal relation (Merriam, 1969). The control of gene activity by elements within egg cytoplasm has been demonstrated and it is not unreasonable to presume that changes exhibited by the sperm nucleus during pronuclear development also allow for its reprograming (Gurdon and Woodland, 1968; Johnson and Rao, 1971). As indicated by Newrock *et al.* (1977):

Egg and sperm are highly specialized cells which differ markedly in packing of their chromatin, and it would seem necessary to restore both genomes to the same state if remodelling processes at subsequent stages of development are to generate the same structural and functional qualities in the two chromosome sets.

Formation of the male pronuclear envelope

Development of the male pronuclear envelope has been studied in an variety of organisms and is similar to the series of events described for nuclear envelope formation in mitotic and meiotic cells (Longo, 1973; Franke, 1974). The timing of the formation of the nuclear envelope that surrounds the dispersed sperm-derived chromatin (male pronuclear envelope) appears to vary. For example, in the sea urchin, *Arbacia*, development of the male pronuclear envelope is initiated during chromatin dispersion. In the surf clam, *Spisula*, formation of the male pronuclear envelope occurs after chromatin dispersion, apparently in concert with the formation of the nuclear envelope of the female pronucleus. In either case morphological events involving the formation of the male pronuclear envelope are similar. Vesicles coalesce along the periphery of the dispersed chromatin and fuse to form elongate cisternae that develop pores (Fig. 10.4). The cisternae fuse to enclose the dispersed chromatin and form a nuclear envelope. Development of a nuclear envelope around *in vitro* dispersing sperm nuclei incubated in egg cytosol via vesicle coalescence has also been described (Lohka and Masui, 1984b). Where portions of the sperm nuclear envelope are incorporated into the structure of the male pronuclear envelope, the elongate cisternae fuse with the sperm-derived membranes, and the latter also become a part of the membranous boundary of the male pronucleus. Regions of the sperm nuclear envelope incorporated into the male pronuclear envelope frequently retain distinctive morphological features that allow their identification at later stages of fertilization or embryonic development, e.g. following the fusion of the male and female pronuclei.

Investigations have been carried out to determine the source(s) of membrane that comprises the male pronuclear envelope (Longo, 1976b). Investigations with the sea urchin, *Arbacia*, demonstrate that portions of the sperm nuclear envelope, specifically the apical and basal regions, are incorporated into the male pronuclear envelope. This raises the question of whether the amount of nuclear envelope present within the incorporated sperm is adequate to completely enclose the dispersed paternally-derived chromatin. In *Arbacia* the amount of membrane

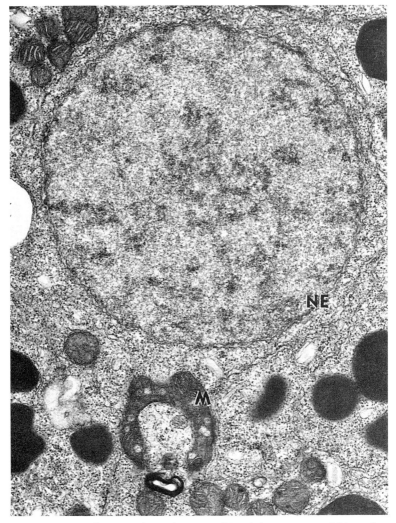

Figure 10.5 Male pronucleus of a sea urchin (*Arbacia*) zygote. NE, nuclear envelope; M, sperm mitochondrion. See Longo (1981b).

present in the sperm nuclear envelope is sufficient to delimit only about 20% of the surface of the male pronucleus.

A likely source of membrane for the formation of the pronuclear envelope, is the endoplasmic reticulum – due to its prevalence, continuity with the nuclear envelope and involvement in nuclear envelope formation and repair in other cells. Investigations of male pronuclear development in centrifuged eggs in which the endoplasmic reticulum is localized to a specific region of the ovum, indicate that the time required to form a male pronuclear envelope is prolonged in areas lacking in and accelerated in areas rich in endoplasmic reticulum (Longo, 1976b). These results strongly suggest that endoplasmic reticulum directly contributes to the formation of the pronuclear envelope.

Although the question of the involvement of membrane biosynthesis in the formation of the pronuclear envelope is a complex one, the following principles seem clear: newly synthesized proteins and lipids appear to be inserted into pre-existing membranes and membrane components are frequently synthesized at or inserted into sites distinct from their ultimate destinations. Membrane biosynthesis in somatic cells has been investigated by measuring the appearance and activity of specific enzymes, the incorporation of labelled precursors into membrane components and the effects of various inhibitory agents on this incorporation. In the case of male pronuclear envelope formation the contribution of membrane biosynthesis is not known with certainty. Analyses with fertilized *Arbacia* eggs demonstrate that the incorporation of labelled leucine into TCA-precipitatable material can be inhibited up to 80% of controls with puromycin during the period of pronuclear development. Electron microscopic observations of puromycin-treated specimens indicate that male pronuclear development is unaffected; the nuclear envelope which forms is morphologically similar to that of the controls. These observations suggest that *de novo* protein synthesis may not be directly involved in pronuclear envelope development.

With the formation of the pronuclear envelope, transformation of the sperm nucleus into a male pronucleus is essentially completed (Fig. 10.5). However, the male pronucleus continues to undergo morphogenetic changes which may include further enlargement, the continuation of chromatin dispersion and the acquisition of intranuclear structures such as nucleoli, annulate lamellae, and aggregations of tubular inclusions.

11

Aspects regulating the development of male pronuclei

The time required for male pronuclear formation varies (Longo, 1973, 1981b): in the sea urchin, *Arbacia*, about 8 min; in the surf clam, *Spisula*, 50–60 min; in the domestic fowl, approximately 25 min; in mammals, 3–4 h. The bases for these time differences have not been determined.

In sea urchins the male pronucleus is much smaller than the female. In other organisms (e.g. mammals), this size difference is less pronounced and may be reversed. If pronuclear migration is inhibited in sea urchin zygotes the male pronucleus will increase in size, often attaining the dimensions of the female pronucleus. These observations suggest that factors responsible for the continued enlargement of the male pronucleus may be present in the zygote cytoplasm after the normal fertilization period and the ultimate size of the male pronucleus may be related to time spent within the zygote cytoplasm. In mammalian zygotes, under similar circumstances, continued enlargement of the male pronucleus is not as obvious.

In eggs which are fertilized at an arrested stage of meiosis the incorporated sperm nucleus transforms into a pronucleus while the maternal chromatin is condensed and engaged in the completion of its meiotic divisions. How the zygote regulates these two very different nuclear activities has not been determined. Observations with *Spisula* zygotes indicate that changes in the sperm nucleus (dispersion) are closely coupled to processes attending meiotic maturation and female pronuclear development (Chen and Longo, 1983; Fig. 11.1). Kinetic analysis of male pronuclear development in *Spisula* demonstrate that male pronuclear enlargement does not proceed at a constant rate but consists of four phases coordinated with major changes in the status of the maternal chromatin. The first phase is a short lag period prior to germinal vesicle breakdown in which the size of the sperm nucleus increases slightly. This is followed by a rapid expansion of the sperm

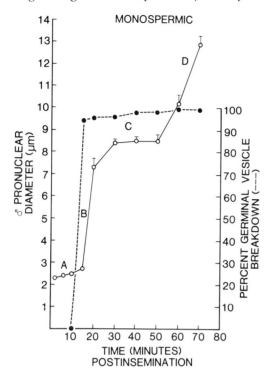

Figure 11.1   Male pronuclear enlargement as determined by its change in diameter vs. time in surf clam (*Spisula*) zygotes. Fertilized eggs were examined at timed intervals to determine the presence or absence of germinal vesicles (- - - -) and the diameter of incorporated sperm nuclei (——). Expansion of the incorporated sperm chromatin is composed of four phases which are correlated with changes in the maternal chromatin: (A) slow, pregerminal vesicle breakdown; (B) rapid, germinal vesicle breakdown; (C) slow, meiotic divisions (polar body formation); and (D) rapid, completion of meiotic maturation/development of the female pronucleus. See Chen and Longo (1983).

nucleus coordinated with germinal vesicle breakdown. With the development of the first meiotic spindle, sperm nuclear enlargement slows dramatically; this lasts until the completion of the meiotic divisions when the developing male pronucleus undergoes a second rapid increase in size that correlates with the female pronuclear development.

The eggs of many animals can be fertilized at an earlier stage than normal, e.g. amphibian and hamster eggs which are normally inseminated at the second metaphase of meiosis can be fertilized at meiotic prophase (germinal vesicle stage). In these cases, however, the incorporated sperm remains essentially unchanged (Dettlaff, Nikitina and Stroeva, 1964; Katagiri, 1974; Usui and Yanagimachi, 1976; Longo, 1978b; Hylander, Anstrom and Summers, 1981). In many instances it is only after germinal vesicle breakdown that the incorporated sperm nucleus undergoes transformations into a male pronucleus. These observations suggest that cytoplasmic factors involving the development of the male pronucleus are established in association with germinal vesicle breakdown. Investigators have shown that factors required for morphogenesis of the sperm nucleus into a male pronucleus are, in fact, derived from substances originating in the germinal vesicle (Katagiri and Moriya, 1976; Thadani, 1979; Balakier and Tarkowski, 1980). Furthermore, in some animals, e.g. sea urchins and mammals (Iwamatsu and Chang, 1972; Longo, 1978b), incorporated sperm nuclei exhibit greater differentiation with increasing oocyte maturation, which suggests that factors required for pronuclear development appear (or are activated) as the egg progresses through meiosis.

Pronuclei that develop from sperm nuclei injected into mature amphibian eggs are capable of DNA synthesis, whereas sperm nuclei injected into immature amphibian eggs neither transform into male pronuclei nor synthesize DNA so long as the germinal vesicle remains intact (Skoblina, 1976; Moriya and Katagiri, 1976). The failure of sperm nuclei to synthesize DNA when injected into immature amphibian eggs may be due neither to a deficiency of DNA polymerase nor to an absence of deoxyribonucleotides but to additional cytoplasmic factors which make the sperm DNA accessible for replication.

A number of studies have indicated that transformation of the sperm nucleus into a male pronucleus may not be entirely dependent upon factors originating with germinal vesicle breakdown. A male pronuclear growth factor is believed to be inactive or absent in *in vitro* matured rabbit, bovine and porcine oocytes (Thibault and Gérard, 1973; Motlik and Fulka, 1974; Trouson, Willadsen and Rowson, 1977). This factor may be formed in surrounding follicular cells stimulated by gonadotropins and then transported to the egg. It has been speculated that *in vitro* culture does not provide a 'natural' environment required for normal cytoplasm development. This suggestion is supported by studies demonstrating that marked qualitative changes occur in protein synthesis during maturation of oocytes *in vitro*. Based on these findings it has been suggested that proteins synthesized during the later stages of maturation are related directly to postmaturational events associated with fertilization and early development. Despite the evidence for a male pronuclear growth factor, other studies have shown that oocytes matured *in vitro* and then transferred to the oviducts of mated animals do have a limited potential for normal embryogenesis.

There is evidence to suggest that factors, instrumental in the transformation of the sperm nuclei, are present in limited quantities. For example, fully formed male pronuclei and undeveloped sperm nuclei may be found within the same cytoplasm of polyspermic ova (Hunter, 1967; Poccia *et al.*, 1978). In these instances, factors responsible for male pronuclear development may have been exhausted or inactivated by developing male pronuclei and, therefore, were absent or unable to influence remaining incorporated sperm nuclei.

Attempts have been made to determine the longevity of cytoplasmic factors that might be involved with the transformation of the sperm nucleus, e.g. how long after insemination is the egg cytoplasm capable of supporting the development of a male pronucleus? Investigations have also considered the possibility that those elements responsible for metamorphosis of the sperm nucleus into a male pronucleus are similar to cell cycle regulators (Longo, 1983). Such an association is not difficult to envisage since many of the events of pronuclear development are similar morphologically to events during mitosis. For example, breakdown of the sperm nuclear envelope, chromatin dispersion and formation of the pronuclear envelope are structurally similar to events occurring at prophase and telophase of mitotically-active cells.

Studies concerning the longevity of the factors regulating pronuclear development primarily involve refertilization or sperm incorporation into somatic cells. Sperm nuclear changes have been examined in the sea urchin (*Arbacia*) and hamster following refertilization to establish the longevity of the factors regulating pronuclear development (Usui and Yanagimachi, 1976; Longo, 1984). It is possible to fuse sperm with *Arbacia* zygotes 20–30 min after insemination and well after pronuclear fusion (approximately 15 min post-fertilization) while the zygote nucleus is undergoing DNA replication. Sperm incorporated into such zygotes undergo male pronuclear development as observed in controls.

Observations of hamster and sea urchin zygotes reinseminated at later stages during fertilization and at different stages of cleavage indicate that regulators of sperm nuclear transformation may have a more general role, relating to factors involved with the cell cycle (Usui and Yanagimachi, 1976; Longo, 1984). For example, pronuclear development does not take place when sperm are incorporated into zygotes at the pronuclear stage, whereas sperm nuclear envelope breakdown and chromatin dispersion take place when sperm are incorporated into zygotes shortly before the first cleavage division, i.e. at prometaphase of mitosis.

Sperm incorporated into cultured somatic cells provide a system to determine whether factors necessary for the transformation of the sperm nucleus into a male pronucleus are present in cells other than eggs. Sperm, phagocytozed by cultured cells, may appear to undergo changes comparable to those observed at fertilization, however, when closely examined, they are degenerative (Phillips *et al.*, 1976). There are instances, where activation of the sperm nucleus (chromatin dispersion) appears to take place within somatic cells (Van Meel and Pearson, 1979). This is accompanied by a shift from the protamine- to the histone-type of basic protein associated with the transforming sperm nucleus. In some instances there is also the induction of RNA and DNA synthesis in the transformed sperm nucleus. These results suggest that factors may be present in somatic cells capable of inducing alterations in incorporated sperm nuclei similar to those occurring in fertilized eggs.

Investigations have examined the specificity of factors involved in pronuclear development, i.e. whether factors responsible for sperm nuclear transformation in the eggs of one species are capable of interacting with the sperm nucleus of a different species to elicit pronuclear development. In those species studied, e.g. mussel (*Mytilus*) ♂ × sea urchin (*Arbacia*) ♀ or crosses between different species of mammal, changes in the sperm nucleus characteristic of pronuclear development do occur (Wu and Chang, 1973; Imai, Niwa and Iritani, 1977; Barros and Herrera, 1977; Longo, 1977). These investigations indicate that factors involving the transformation of the sperm nucleus into a male pronucleus are not species specific and are capable of interacting with sperm nuclei of evolutionary divergent organisms. The extent to which such interspecies combinations accomplish all the events of pronuclear formation has not been established. The male pronuclei of naturally fertilized *Mytilus* and *Arbacia* eggs differ when compared morphologically. Interestingly, in the cross of *Mytilus* ♂ × *Arbacia* ♀, the male pronuclei resemble those that develop from *Arbacia* sperm, suggesting that cytoplasmic factors determine the form of the pronucleus.

Experiments in which sperm nuclei have been microinjected into eggs demonstrate that the structural integrity of the spermatozoon and the normal processes of gamete fusion and incorporation are not necessary prerequisites for pronuclear development (Skoblina, 1976; Moriya and Katagiri, 1976; Uehara and Yanagimachi, 1976). Moreover, frozen-thawed and freeze-dried human sperm injected into hamster eggs can develop into structures morphologically resembling male pronuclei (Uehara and Yanagimachi, 1977).

The elevation of intracellular pH that occurs in the ova of some organisms at fertilization is necessary for egg activation and development of the male pronucleus (Chambers, 1976; Carron and Longo, 1980). In sea urchins, if alkalinization of the egg cytoplasm is blocked, male pronuclear development is reversibly inhibited which suggests that pronuclear formation is dependent upon zygote cytoplasmic alkalinization. Experiments with somatic cells also indicate that intracellular pH modulation influences nuclear structure.

12

Fates of incorporated sperm mitochondria, flagella and perinuclear structures

In most animals, incorporation of the sperm nucleus is accompanied by the entry of the sperm mitochondria and components of the sperm flagellum, the axonemal complex (Longo, 1973; Fig. 12.1). Structural alterations of the sperm mitochondria have been observed in various mammalian species, including swelling and loss of cristae, that are believed to be indicative of degeneration. In the mouse, the sperm mitochondria begin to degenerate during the early stages of fertilization, whereas in the rat changes are not noted until cleavage. Analysis of mitochondrial DNAs of hybrid amphibian embryos derived from *Xenopus laevis* and *Xenopus mülleri* indicate that mitochondria are maternally derived (Dawid and Blackler, 1972).

In the sea urchin, *Arbacia*, changes observed in sperm mitochondria during fertilization are not consistent. Some zygotes may exhibit a decrease in electron opacity, a loss of cristae and some swelling, while in others, little or no change is apparent. Recognizable sperm mitochondria have been observed juxtaposed to the first mitotic apparatus, approximately 60 min following insemination. Sperm mitochondria appear to decrease in size and become structurally similar to those maternally derived during the early stages of fertilization in the mussel, *Mytilus*. Due to these changes, they become indistinguishable from maternally-derived mitochondria in the surrounding cytoplasm.

In ascidians, the spermatozoon, when it contacts the egg chorion loses its mitochondrion within 2 min of insemination. The mitochondria slide along the tail to its tip and are released (Lambert and Lambert, 1981). Hence, in this case the sperm mitochondria are not incorporated into the egg. This process is calcium-dependent and can be induced by egg water, elevated pH and low external sodium. The reaction is

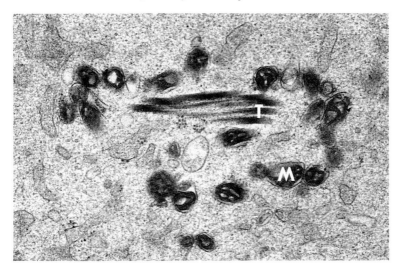

Figure 12.1   Incorporated sperm mitochondria (M) and portion of the sperm tail (T) in a rabbit zygote. See Longo (1976c).

accompanied by a drop in extracellular pH – suggestive of a release of protons from the sperm. The release of the mitochondria may be necessary for fertilization because of steric factors, so allowing the sperm to pass through the chorion, i.e. it may provide the push for sperm incorporation.

The incorporated sperm axoneme has been observed in mammalian zygotes, and has been followed by light and electron microscopy during cleavage. The external fibers and axonemal complex are observed in mouse, rat, rabbit, hamster and human zygotes (Zamboni, 1971). In the rat, very little change is observed in the sperm axoneme up to the second division, following which it disappears (Szollosi, 1965). The incorporated sperm axoneme of the sea urchin, *Arbacia*, appears to retain its structural integrity up to the first cleavage division, its fate at later stages of development is unknown (Longo, 1973). In the mussel, *Mytilus*, portions of the sperm axoneme are located adjacent to the sperm nucleus immediately after incorporation. Subsequently, the sperm axoneme is not observed although long segments of microtubules, believed to be derived from this structure, may be found in association with the reorganizing sperm nucleus. Short segments of incorporated sperm axoneme have been observed in the surf clam, *Spisula*, and it is possible that little of this structure is taken into the egg upon insemination.

The incorporated post-acrosomal complex of mouse zygotes has been observed up to the period of male pronuclear development (Stefanini, Oura and Zamboni, 1969). The fate of this structure has not been elucidated. Acid phosphatase activity has been demonstrated in the post-acrosomal region of the mammalian spermatozoon which may modify egg cytoplasmic components following sperm incorporation.

13

Development of the sperm aster and pronuclear migration

Development of the male pronucleus is usually restricted to the periphery of the zygote. However, in some cases, the dispersing sperm chromatin may begin to migrate prior to its development into a male pronucleus. The direction or path taken by the male pronucleus was the subject of investigations among early cytologists, who distinguished two components: (a) a penetration path which is approximately vertical to the zygote's surface and (b) a copulation path which brings the pronuclei into close association (Wilson, 1925; Fig. 13.1). Accompanying the male pronucleus during these movements is the sperm aster (Fig. 13.2).

Sperm aster development has been studied in the eggs of sea urchins, molluscs and mammals (Longo, 1973, 1976c; Schatten, 1984). Sperm aster formation does not occur in inseminated *Arbacia* oocytes and this lack of development may be due to the absence of a cytoplasmic constitutent(s) which acts in conjunction with the incorporated sperm components to initiate its morphogenesis. Formation of the sperm aster in the various organisms studied to date appears to differ temporally with respect to male pronuclear development. For example, in *Arbacia*, morphogenesis of the sperm aster may begin during the early stages of chromatin dispersion but usually occurs during the formation of the male pronuclear envelope, about 6 min postinsemination. In the surf clam (*Spisula*) sperm aster formation commences following meiosis of the maternal chromatin and during the formation of the male pronuclear envelope, approximately 50 min postinsemination.

One of the earliest indications of sperm aster development is the dissociation of the centrioles from the sperm axoneme. In the molluscs, *Mytilus* and *Spisula*, the distal and proximal centrioles move into the

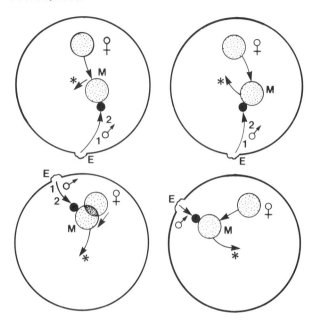

Figure 13.1 Paths of the male and female pronuclei during their movements to each other and to the center of the zygote. E, entrance point of the spermatozoon; ♀ original position of the female pronucleus; ♂, path of the male pronucleus; M, meeting point of the pronuclei; *, path taken by the zygote nucleus to become located within the center of the zygote. The track taken by the male pronucleus may be resolved into two components, a penetration path (1), nearly vertical to the egg surface and a copulation-path (2) along which the male pronucleus moves towards the point of association with the female pronucleus.

cytoplasmic area in advance of the developing male pronucleus. In the sea urchin, *Arbacia*, only the distal centriole moves from the reorganizing sperm chromatin, while the proximal centriole is retained within an invagination of the male pronucleus, the centriolar fossa.

Observations of rabbit zygotes did not disclose centrioles in the centrosphere of the developing sperm asters (Longo, 1976c). In *Arbacia*, *Mytilus* and *Spisula*, both the distal and proximal sperm centrioles appear to function as organization centers for the assembly of sperm aster microtubules. Initially, large quantities of endoplasmic reticulum and microtubules aggregate in the pericentriolar region and form a dense matrix, referred to as the centrosphere. The formation of the centrosphere and its subsequent growth tend to exclude components, such as mitochondria, yolk and lipid bodies, which become confined to its periphery. Radiating from the centrosphere are fascicles of micro-

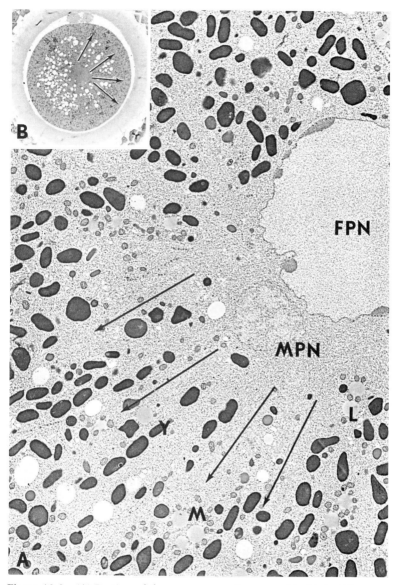

Figure 13.2 (A) Portion of the sperm aster of a sea urchin (*Arbacia*) zygote showing radiating fasicles (arrows) containing microtubules and endoplasmic reticulum (not depicted because of the low magnification) which are separated by columns containing yolk (Y) and lipid (L) bodies and mitochondria (M). MPN and FPN, male and female pronuclei. (B) Sperm aster (arrows) of a rabbit zygote. See Longo (1973).

tubules interspersed among elements of endoplasmic reticulum. These radiating bundles of tubular-lamellar components are separated by clusters of yolk bodies, mitochondria and lipid droplets (Fig. 13.2).

During later stages of development the sperm aster enlarges, as a result of an accumulation of microtubules and endoplasmic reticulum. The morphology of the sperm aster is similar to the asters that comprise the mitotic and meiotic apparatus. However, they appear to differ in that the pericentriolar region of the meiotic and mitotic asters is frequently larger and composed of a dense matrix of fine-textured material.

The involvement of microtubules in pronuclear migration is indicated by studies in which chemical and physical methods known to destroy microtubular structures also prevent the movements and association of the pronuclei (Zimmerman and Zimmerman, 1967; Longo, 1976c; Schatten, 1984). The absence of microfilaments and an insensitivity of pronuclear movements to cytochalasin B suggest that pronuclear migration does not involve actin.

Based largely on observations of the sea urchin egg, in which the oocyte centrioles disappear some time during maturation, it was concluded that the egg possesses all the elements necessary for development except centrioles (Mazia, 1961). The spermatozoon, on the other hand, possesses centrioles but lacks the necessary medium in which to function. Hence, it was inferred that the sperm supplies the division center normally responsible for the cleavage of the zygote. This was first outlined by Boveri (see Wilson, 1925) who postulated that the incorporation of sperm centrioles was a prerequisite for the organization of the cleavage spindle. Numerous investigations, however, indicate that this may not be true in all instances and therefore challenge the concept of the exclusive paternal derivation of the centrioles.

1. Fertilization studies in a number of animals document the participation of egg-derived centrioles in the organization of the first cleavage spindle.
2. The production of cytasters containing centrioles in parthenogenetically activated eggs of the sea urchin (Dirksen, 1961).
3. The formation of cilia by those organisms which normally develop parthenogenetically.
4. The lack of participation of centrioles in the formation of the cleavage spindle in mammalian embryos until at least the fourth cleavage stage (Szollosi, Calarco and Donahue, 1972).
5. Kinetosome formation during ciliogenesis (Sorokin, 1968; Steinman, 1968).

There exist a number of ways in which the centrioles are brought into or are formed by the zygote. Considering the apparent contribution of maternal centrioles to the mitotic apparatus in those forms which have been studied at the ultrastructural level of investigation and the absence of centrioles in the egg and zygote of some mammals, further examination is needed to clarify our understanding of the origin and morphogenesis of centrioles during development.

Pronuclear migration involves those processes which are responsible for the movement of the male and female pronuclei from their site of formation to the region where they become associated. The end result is the juxtaposition of the pronuclei. How pronuclear movements are brought about has been the subject of considerable speculation.

1. Migration of the male pronucleus may be due to the elongation of sperm aster components; that the enlarging asters of dispermic eggs are pushed apart supports such a scheme.
2. Dense aggregations of microtubules in the cytoplasm between the male and female pronuclei have been observed in brine shrimp (*Artemia*) zygotes (Anteanis, Fautrez-Firtefyn and Fautrez, 1967). This array of microtubules may be instrumental in drawing the two pronuclei together.
3. The astral rays in eggs of the gall midge adhere to peripheral structures and exert a tractive force that pulls the aster through the ooplasm (Wolf, 1978).
4. Sperm aster elements may establish a field of streaming cytoplasm which is ultimately responsible for moving the pronuclei together (Chambers, 1939).

Aspects regulating the movement of the female pronucleus are unknown and it has not been clearly established whether or not they are related to mechanism(s) responsible for male pronuclear migration. In many zygotes, the female pronucleus is formed near where the pronuclei become associated and therefore, it may undergo relatively little movement. In some species, the female pronucleus is associated with an aster-like structure which may function in the manner hypothesized for the sperm aster (Longo, 1973). Movement of the female pronucleus however, can apparently occur in the absence of a sperm aster. For example, artificially activated sea urchin eggs are distinguished by the migration of the female pronucleus to the center of the egg. In *Lytechinus* eggs there is a radially oriented population of microtubules which is not associated with the sperm aster (Mar, 1980). The microtubules originate in the egg cortex and elongate radially into the center of the egg. Elongation of the radially oriented population of microtubules is associated with the centrad migration of the male and female pronuclei in fertilized eggs or the female pronucleus in artificially activated ova (Harris, 1979; Harris, Osborn and Weber, 1980).

On the basis of fluorescent studies of cytoskeletal components of sea urchin zygotes Schatten (1984) suggested that the initial assembly of microtubules on sperm centrioles pushes the male pronucleus to the center of the egg. This is believed to be a steric effect resulting from the elongation of these elements. When sperm aster microtubules elongate to contact the surface of the female pronucleus, a swift migration of the female pronucleus to the male pronucleus occurs. During this movement, the sperm aster becomes asymmetric. It is possible that disassembly of microtubules that interconnect the female and male pronuclei generate the force for movement of the female pronucleus. Following the migration of the female pronucleus to the sperm aster, the adjacent pronuclei move to the center of the zygote. This movement is believed to be due to an elongation of the sperm aster microtubules.

Association of the male and female pronuclei: The concluding events of fertilization

The end result of the migration of the pronuclei is their juxtaposition. When adjacent, the pronuclei become associated in a characteristic fashion to establish the genome of the embryo and conclude the process of fertilization (Fig. 14.1). Early studies of fertilization indicated that the association of the male and female pronuclei may take essentially two forms.

1. Both the male and female pronuclei may give rise to a group of chromosomes for the ensuing cleavage division. In this form, there is an intermixing of the maternal and paternal chromosomes without fusion of the pronuclei. Eggs demonstrating this series of events are referred to as possessing the *Ascaris*-type of fertilization (Wilson, 1925).
2. Fusion of the pronuclei to produce a single zygote nucleus (pronuclear fusion). Eggs exhibiting such a process are said to have the sea urchin-type of fertilization.

Ascaris-type of fertilization

The events associated with the Ascaris-type of fertilization have been studied in the parasitic intestinal round worm *Ascaris*, in the molluscs *Spisula* and *Mytilus* and various mammals (Zamboni, 1971; Longo, 1973). Eggs demonstrating this type of pronuclear association are inseminated prior to the completion of meiosis, i.e. at meiotic prophase (germinal vesicle stage), metaphase I or metaphase II. Differences have been noted in each of the organisms studied and there is evidence that

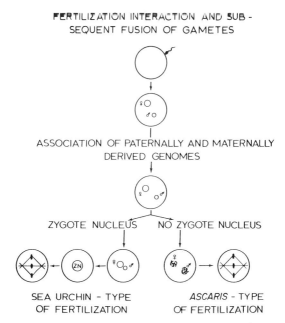

Figure 14.1   Schematic representation of the principal ways in which the maternally and paternally derived genomes become associated during fertilization. ZN, zygote nucleus; ♂, male pronucleus/chromosomes; ♀, female pronucleus/chromosomes. Reproduced with permission from Longo (1973).

these variations may be related to the meiotic stage of the egg at insemination (Longo, 1973).

At the time of their association the male and female pronuclei of zygotes having the *Ascaris*-type of fertilization are large spheroids (Figs 14.2, 14.3). The pronuclei may become closely apposed and form nucleoplasmic projections that interdigitate. Concomitantly, two asters become situated to either side of the associated pronuclei and establish what will become the poles of the mitotic spindle for first cleavage. During this period condensing chromosomes appear within the male and female pronuclei and the pronuclear envelopes breakdown (Fig. 14.4). Ultimately, vesicles and elongated cisternae form that initially outline the condensing chromosomes but are eventually scattered within the cytoplasm. Following the disruption of the pronuclear envelopes, the chromosomal groups from each parent intermix on what will constitute the metaphase plate of the first mitotic apparatus. Microtubules become associated with the chromosomes and course to regions of the asters (Fig. 14.5). During this process a zygote nucleus is

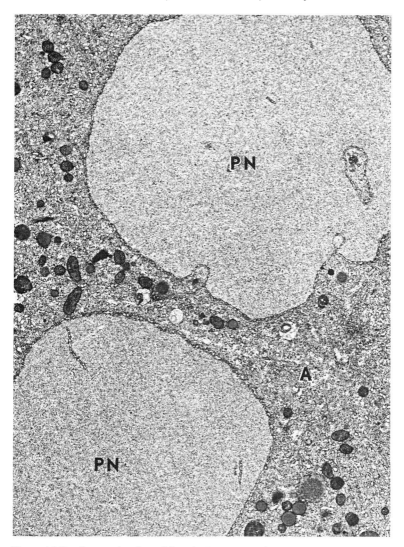

Figure 14.2 Apposed male and female (PN) pronuclei associated with an aster (A) in a surf clam (*Spisula*) zygote. See Longo and Anderson (1970b).

not formed, i.e. the pronuclei do not fuse. The maternal and paternal genomes are associated within a single nucleus for the first time in the blastomeres of the two-cell stage (Fig. 14.6.).

Many of the morphological aspects characteristic of the *Ascaris*-type of fertilization are in fact early events normally associated with mitosis,

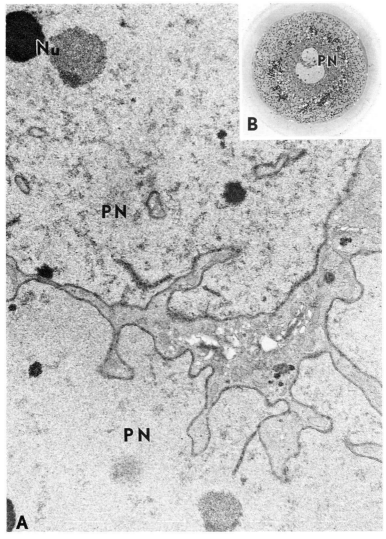

Figure 14.3   Associated pronuclei (PN) in a rabbit zygote. Nu, nucleoli. See Longo and Anderson (1969b).

i.e. prophase. Therefore, in these forms, the later stages of fertilization also encompass early events of mitosis. Moreover, the eggs thus far known to demonstrate the *Ascaris*-type of fertilization are inseminated at an arrested stage of meiosis. Consequently, fertilization in these forms involves:

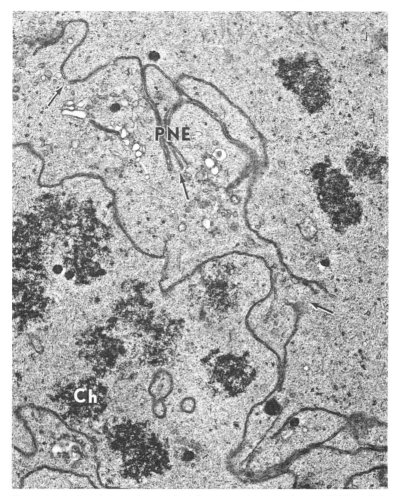

Figure 14.4 Breakdown (arrows) of the pronuclear envelopes (PNE) and the condensation of chromatin (Ch) in a rabbit zygote. See Longo and Anderson (1969b).

1. The continuation of meiosis of the maternal chromatin.
2. The events of fertilization *per se*, e.g. sperm incorporation and pronuclear development.
3. The early stages of mitosis, e.g. prophase to metaphase.

Sea urchin-type of fertilization

The sea urchin-type of fertilization is observed in eggs that are

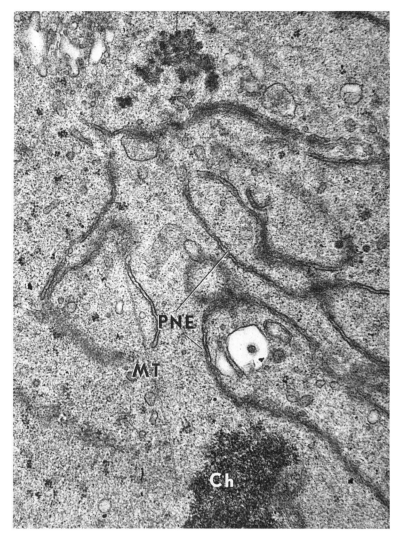

Figure 14.5 Section through a rabbit zygote during an advanced stage of the breakdown of the male and female pronuclear envelopes (PNE). Ch, chromosomes associated with spindle microtubules (MT).

inseminated after meiotic maturation. It involves a fusion of the nuclear envelopes of the male and female pronuclei (pronuclear fusion) and results in the formation of the zygote nucleus (Fig. 14.7). This process has been studied in greatest detail in *Lytechinus* and in *Arbacia* (Wilson, 1925; Longo, 1973). A specific site of contact and fusion does not exist

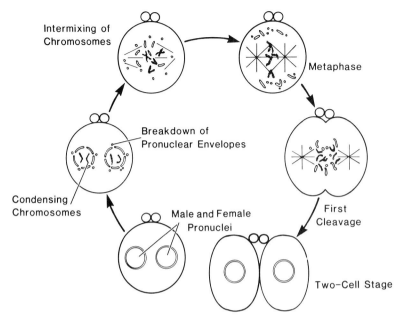

Figure 14.6 Diagrammatic representation of pronuclear association in eggs undergoing the *Ascaris*-type of fertilization.

along the periphery of the pronuclei and pronuclear fusion appears to occur randomly along the surfaces of both pronuclei. Following contact of the pronuclei, the outer membranes of both pronuclear envelopes fuse, so that the inner membranes of both pronuclei are brought into apposition (Figs. 14.8, 14.9). Subsequently, the inner membranes of the male and female pronuclear envelope fuse and form an internuclear bridge joining the former pronuclei (Fig. 14.10). How this process is regulated has not been established; however, fusion of somatic cell nuclei *in vitro* appears to be modulated by guanosine triphosphate (Paiement, Beaufay and Godelaine, 1980).

Initially the internuclear bridge connecting the former male and female pronuclei is small; it gradually increases in diameter, eventually yielding a spheroid zygote nucleus with a slight protuberance that represents the former male pronucleus (Fig. 14.11). The paternal chromatin is seen at the fusion site as an electron dense mass; eventually this material diffuses throughout the zygote nucleus and becomes indistinguishable from that of the female (Fig. 14.11).

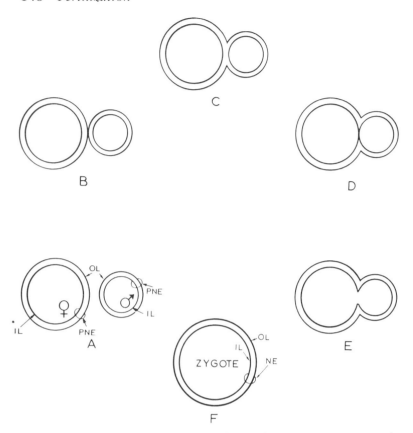

Figure 14.7   Schematic representation of pronuclear association in eggs undergoing the sea urchin-type of fertilization (pronuclear fusion). (A) Apposition of the male (♂) and female (♀) pronuclei. (B) Contact of the outer laminae (OL) of the male and female pronuclear envelopes (PNE). (C) Fusion of the outer laminae and apposition of the inner laminae (IL). (D) Contact of the inner laminae of the pronuclear envelopes. (E) Fusion of the inner laminae of the pronuclear envelopes. (F) Zygote nucleus with a continuous nuclear envelope (NE). Reproduced with permission from Longo (1973).

Comparison of the *Ascaris*- and the sea urchin- types of fertilization

The differences between the sea urchin-type of fertilization and the *Ascaris*-type of fertilization appear to be determined, at least in part, by the time interval between the entrance of the spermatozoon and the association of the pronuclei. By artificially prolonging this period the

Figure 14.8   (*top*) Initial stage in the fusion of the male (MPN) and female (FPN) pronuclei in the sea urchin, *Arbacia* (arrow denotes site). See Longo and Anderson (1968).

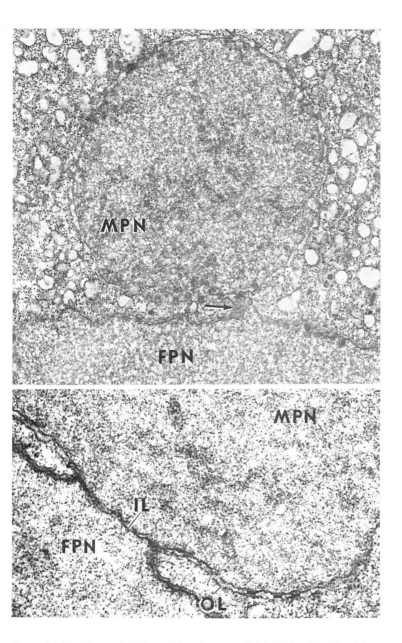

Figure 14.9. (*bottom*) Male and female pronuclei (MPN and FPN) of the sea urchin, *Arbacia* following fusion of the outer laminae (OL) of the male and female pronuclear envelope. The inner laminae (IL) of the pronuclear envelopes are parallel to one another. See Longo (1973).

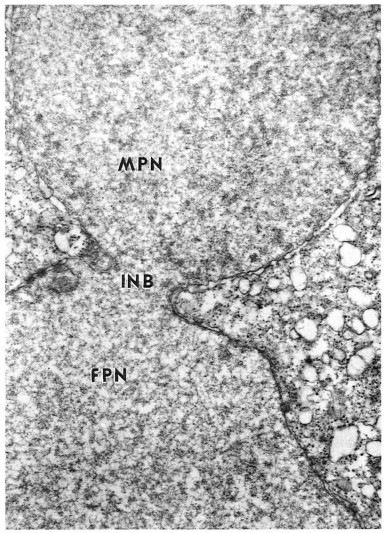

Figure 14.10   Result of the fusion of the inner laminae of the male and female pronuclear envelopes in a sea urchin (*Arbacia*) zygote. An internuclear bridge (INB) connects the former male (MPN) and female (FPN) pronuclei. See Longo and Anderson (1968).

sea urchin-type of fertilization may take on the character of the *Ascaris*-type (Wilson, 1925; Longo and Anderson, 1970c,d). For example, a lengthy delay of pronuclear fusion in *Arbacia* results in an enlargement of the male pronucleus and the acquisition of structural

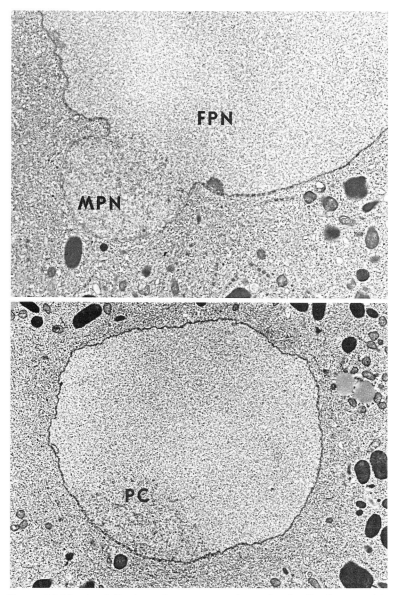

Figure 14.11 (*top*) Portion of a zygote nucleus of a fertilized sea urchi (*Arbacia*) egg containing a protuberance that was formerly the male pronucleu (MPN). FPN, former female pronucleus.

Figure 14.12 (*bottom*) Zygote nucleus of sea urchin, *Arbacia* containing dense paternally derived chromatin (PC) at the region which marks the site of pronuclear fusion. Reproduced with permission from Longo and Anderson (1968).

features normally associated with the female pronucleus. In some instances, the male pronucleus may undergo a similar series of events observed in the molluscs *Spisula* and *Mytilus* and some mammals, i.e. vesiculation of the pronuclear envelopes and chromatin condensation.

Wilson (1925) indicated that due to the varying relation between the time of polar body formation and sperm incorporation, the sea urchin- and the *Ascaris*-type of fertilization may represent extremes connected by a series of intermediate forms that have been described (Longo, 1973). Why zygotes exhibit a varied series of pronuclear events, modulated in a characteristic fashion, is an unexplained and interesting facet of fertilization. However, aside from the differences and similarities exhibited in fertilized eggs from various organisms, the fact of primary importance is the bringing together of two originally separated lines of heredity, thereby establishing the genetic composition of the embryo.

References

Ahuja, K.K. (1982) Fertilization studies in the hamster: The role of cell-surface carbohydrates. *Exp. Cell Res.*, **140**, 353–62.

Aketa, K. (1973) Physiological studies on the sperm surface component responsible for sperm–egg bonding in sea urchin fertilization. *Exp. Cell Res.*, **80**, 439–41.

Aketa, K. and Tsuzuki, H. (1968) Sperm-binding capacity of the S-S reduced protein of the vitelline membrane of the sea urchin egg. *Exp. Cell Res.*, **50**, 675–6.

Aketa, K. and Onitake, K. (1969) Effect on fertilization of antiserum against sperm-binding from homo- and heterologous sea urchin egg surfaces. *Exp. Cell Res.*, **56**, 84–6.

Aketa, K. and Ohta, T. (1977) When do sperm of the sea urchin, *Pseudocentrotus depressus*, undergo the acrosome reaction at fertilization? *Dev. Biol.*, **61**, 366–72.

Anderson, E. (1968) Oocyte differentiation in the sea urchin, *Arbacia punctulata*, with particular reference to the cortical granules and their participation in the cortical reaction. *J. Cell Biol.*, **37**, 514–39.

Anderson, W.A. and Eckberg, W.R. (1983) A cytological analysis of fertilization in *Chaetopterus1 pergamentaceus*. *Biol. Bull.*, **165**, 110–18.

Anteanis, A., Fautrez-Firtefyn, N. and Fautrez, J. (1967) L'accolement des pronuclei de l'seuf *d'Artemia salina*. *J. Ultrastruct. Res.*, **20**, 206–10.

Austin, C.R. (1951) Observations on the penetration of sperm into the mammalian egg. *Austral. J. Sci. Res.*, (B) **4**, 581–96.

Austin, C.R. (1961) *The Mammalian Egg*, Blackwell Scientific Publications, Oxford, England.

Austin, C.R. (1964) Sperm centrioles and their role in fertilization. *Proc. Int. Congr. Anim. Reprod.*, 7, 302.

Austin, C.R. (1965) *Fertilization*, Prentice Hall, Englewood Cliffs, New Jersey.

Austin, C.R. and Bishop, M.W.H. (1957) Fertilization in mammals. *Biol. Rev.*, **32**, 296–349.

Azarnia, R. and Chambers, E.L. (1976) The role of divalent cations in activation of the sea urchin egg. I. Effect of fertilization on divalent cation content. *J. Exp. Zool.*, **198**, 65–78.

Babcock, D.F., Singh, J.P. and Lardy, H.A. (1979) Alteration of membrane permeability to calcium ions during maturation of bovine spermatozoa. *Dev. Biol.*, **69**, 85–93.

Baccetti, B. and Afzelius, B. (1976) *The Biology of the Sperm Cell*, Karger, New York.

Baker, P.F. and Whitaker, M.J. (1978) Influence of ATP and calcium on the cortical reaction in sea urchin eggs. *Nature*, **276**, 513–15.

Baker, P.F. and Whitaker, M.J. (1979) Trifluoroperazine inhibits exocytosis in sea urchin eggs. *J. Physiol.*, **298**, 55.

Balakier, H. and Czolowska, R. (1977) Cytoplasmic control of nuclear maturation in mouse oocytes. *Exp. Cell Res.*, **110**, 466–9.

Balakier, H. and Tarkowski, A.K. (1980) The role of germinal vesicle karyoplasm in the development of male pronucleus in the mouse. *Exp. Cell Res.*, **128**, 79–85.

Barros, C. and Herrera, E. (1977) Ultrastructural observations of the incorporation of guinea-pig spermatozoa into zona-free hamster oocytes. *J. Reprod. Fert.*, **49**, 47–50.

Barros, C., Bedford, J.M., Franklin, L.E. and Austin, C.R. (1967) Membrane vesiculation as a feature of the mammalian acrosome reaction. *J. Cell Biol.*, **34**, C1–C5.

Barry, J.M. and Merriam, R.W. (1972) Swelling of hen erythrocyte nuclei in cytoplasm from *Xenopus* eggs. *Exp. Cell Res.*, **71**, 90–96.

Bedford, J.M. (1970) The saga of mammalian sperm from ejaculation to syngamy, in *Mammalian Reproduction* (eds H. Gibian and E.J. Plotz) Springer-Verlag, New York, pp. 124–82.

Bedford, J.M. and Cooper, G.W. (1978) Membrane fusion events in the fertilization of vertebrate eggs, in *Cell Surface Reviews*, Volume 5 (eds G. Poste and G.L. Nicolson), Elsevier North-Holland, Amsterdam, pp. 65–125.

Begg, D.A., Rebhun, L.I. and Hyatt, H. (1982) Structural organization of actin in the sea urchin egg cortex: Microvillar elongation in the absence of actin filament bundle formation. *J. Cell Biol.*, **93**, 24–32.

Bell, E. and Reeder, R. (1967) The effect of fertilization on protein synthesis in the egg of the surf-clam *Spisula solidissima*. *Biochim. Biophys. Acta*, **142**, 500–11.

Bennett, J. and Mazia, D. (1981a) Interspecific fusion of sea urchin eggs. Surface events and cytoplasmic mixing. *Exp. Cell Res.*, **131**, 197–207.

Bennett, J. and Mazia, D. (1981b) Fusion of fertilized and unfertilized sea urchin eggs. Maintenance of cell surface integrity. *Exp. Cell Res.* **134**, 494–8.

Berridge, M.J. and Irvine, R.F. (1984) Inositol trisphosphate, a novel second messenger in cellular signal transduction. *Nature* **312**, 315–21.

Bleil, J.D. and Wassarman, P.M. (1980a) Structure and function of the zona pellucida: Identification and characterization of the proteins of the mouse oocyte's zona pellucida. *Dev. Biol.*, **76**, 185–202.

Bleil, J. and Wassarman, P. (1980b) Mammalian sperm–egg interaction: Identification of a glyprotein in mouse zonae pellucidae possessing receptor activity for sperm. *Cell*, **20**, 873–82.

Bleil, J. and Wassarman, P. (1983) Sperm–egg interactions in the mouse: Sequence of events and induction of the acrosome reaction by a zona pellucida glyprotein. *Dev. Biol.*, **95**, 317–24.

Bloch, D.P. (1969) A catalog of sperm histones. *Genetics*, Suppl. **61**, 93–111.

Bloch, D.P. and Hew, H.Y.C. (1960) Changes in nuclear histones during fertilization, and early embryonic development in the pulmonate snail, *Helix aspersa*. *J. Biophys. Biochem. Cytol.*, **8**, 69–81.

Brachet, J., Decroly, M., Ficq, A. and Quertier, J. (1963) Ribonucleic acid metabolism in unfertilized and fertilized sea-urchin eggs. *Biochim. Biophys. Acta* **72**, 660–2.

Bradley, M.P. and Forrester, I.T. (1980) A [Ca^{2+} + Mg^{2+}] -ATPase and active Ca^{2+} transport in the plasma membranes isolated from ram sperm flagella. *Cell Calcium*, **1**, 381–90.

Brandhorst, B.P. (1976) Two-dimensional gel patterns of protein synthesis before and after fertilization of sea urchin eggs. *Dev. Biol.*, **52**, 310–17.

Brandis, J.W. and Raff, R.A. (1978) Translation of oogenetic mRNA in sea urchin eggs and early embryos. Demonstration of a change in translational efficiency following fertilization. *Dev. Biol.* **67**, 99–113.

Bryan, J. (1970a) The isolation of a major structural element of the sea urchin fertilization membrane. *J. Cell Biol.*, **44**, 635–44.

Bryan, J. (1970b) On the reconstitution of the crystalline components of the sea urchin fertilization membrane. *J. Cell Biol.*, **45**, 606–14.

Bryan, J. and Kane, R.E. (1982) Actin gelatin in sea urchin egg extracts. *Meth. Cell Biol.*, **25**, 175–99.

Burnside, B., Kozak, C. and Kafatos, F.C. (1973) Tubulin determination by an isotope dilution-vinblastine precipitation method. The tubulin content of *Spisula* eggs and embryos. *J. Cell Biol.*, **59**, 755–62.

Calarco, P.G., Donahue, R.P. and Szollosi, D. (1972) Germinal vesicle breakdown in the mouse oocyte. *J. Cell Sci.*, **10**, 369–85.

Calvin, H.I. and Bedford, J.M. (1971) Formation of disulphide bonds in the nucleus and accessory structures of mammalian spermatozoa during maturation in the epididymis. *J. Reprod. Fert.*, Suppl. **13**, 65–75.

Campanella, C. (1975) The site of spermatozoon entrance in the unfertilized egg of *Discoglossus pictus* (Anura): An electron microscope study. *Biol. Reprod.*, **12**, 439–47.

Campanella, C., Andreuccetti, P., Taddei, C. and Talevi, R. (1984) The modifications of cortical endoplasmic reticulum during *in vitro* maturation of *Xenopus laevis* oocytes and its involvement in cortical granule exocytosis. *J. Exp. Zool.*, **229**, 283–93.

Campisi, J. and Scandella, C.J. (1978) Fertilization-induced changes in membrane fluidity of sea urchin eggs. *Science*, **199**, 1336–7.

Campisi, J. and Scandella, C.J. (1980a) Bulk membrane fluidity increases after fertilization or partial activation of sea urchin eggs. *J. Biol. Chem.*, **255**, 5411–19.

Campisi, J. and Scandella, C.J. (1980b) Calcium-induced decrease in membrane fluidity of sea urchin egg cortex after fertilization. *Nature*, **286**, 185–6.

Carre, D. and Sardet, C. (1981) Sperm chemotaxis in Siphonophores. *Biol. Cell*, **40**, 119–28.

Carroll, E.J. and Epel, D. (1975) Isolation and biological activity of the proteases released by sea urchin eggs following fertilization. *Dev. Biol.*, **44**, 22–32.

Carroll, A.G. and Ozaki, H. (1979) Changes in the histones of the sea urchin *Strongylocentrotus purpuratus* at fertilization. *Exp. Cell Res.*, **119**, 307–15.

Carron, C.P. and Longo, F.J. (1980) Relation of intracellular pH and pronuclear development in the sea urchin, *Arbacia punctulata. Dev. Biol.*, 79, 478–87.

Carron, C.P. and Longo, F.J. (1982) Relation of cytoplasmic alkalinization to microvillar elongation and microfilament formation in the sea urchin egg. *Dev. Biol.*, 89, 128–37.

Carron, C.P. and Longo, F.J. (1983) Filipin/sterol complexes in fertilized and unfertilized sea urchin egg membranes. *Dev. Biol.*, 99, 482–8.

Carron, C.P. and Longo, F.J. (1984) Pinocytosis in fertilized sea urchin (*Arbacia punctulata*) eggs. *J. Exp. Zool.*, 231, 413–22.

Cartaud, A., Boyer, J. and Ozon, R. (1984) Calcium sequestering activities of reticulum vesicles from *Xenopus laevis* oocytes. *Exp. Cell Res.*, 155, 565–74.

Chambers, E.L. (1939) The movement of the egg nucleus in relation to the sperm aster in the echinoderm egg. *J. Exp. Biol.*, 16, 409–24.

Chambers, E.L. (1976) Na is essential for activation of the inseminated sea urchin egg. *J. Exp. Zool.*, 197, 149–54.

Chambers, E.L. (1980) Fertilization and cleavage of eggs of the sea urchin *Lytechinus variegatus* in Ca^{2+} -free sea water. *Eur. J. Cell Biol.*, 22, 476.

Chambers, E.L. and Hinkley, R.E. (1979) Non-propagated cortical reactions induced by the divalent ionophore A-23187 in eggs of the sea urchin, *Lytechinus variegatus. Exp. Cell Res.*, 124, 441–6.

Chambers, E.L., Pressman B.C. and Rose, B. (1974) The activation of sea urchin eggs by the divalent ionophores A23187 and X-537A. *Biochem. Biophys. Res.*, 60, 126–32.

Chandler, D.E. and Heuser, J. (1979) Membrane fusion during secretion: Cortical granule exocytosis in sea urchin eggs as studied by quick-freezing and freeze-fracture. *J. Cell Biol.*, 83, 91–108.

Chandler, D.E. and Heuser, J. (1980) The vitelline layer of the sea urchin egg and its modification during fertilization. *J. Cell Biol.*, 84, 618–32.

Chandler, D.E. and Heuser, J. (1981) Postfertilization growth of microvilli in the sea urchin egg: New views from eggs that have been quick-frozen, freeze-fractured and deeply etched. *Dev. Biol.*, 92, 393–400.

Chang, M.C. (1951) The fertilizing capacity of spermatozoa deposited into the fallopian tubes. *Nature*, 168, 697–8.

Channing, C.P., Anderson, L.D., Hoover, D.J., Kolena, J., Osteen, K.G., Pomerantz, S.H. and Tanabe, K. (1982) The role of non-steroidal regulators in control of oocyte and follicular maturation. *Rec. Prog. Horm. Res.*, 38, 331–408.

Charbonneau, M. and Grey, R.D. (1984) The onset of activation responsiveness during maturation coincides with the formation of the cortical endoplasmic reticulum in oocytes of *Xenopus laevis. Dev. Biol.*, 102, 90–7.

Chen, D.Y. and Longo, F.J. (1983) Sperm nuclear dispersion coordinate with meiotic maturation in fertilized *Spisula solidissima* eggs. *Dev. Biol.*, 99, 217–24.

Cicirelli, M.F., Robinson, K.R. and Smith, L.D. (1983) Internal pH of *Xenopus* oocytes: A study of the mechanism and role of pH changes during meiotic maturation. *Dev. Biol.*, 100, 133–46.

Citkowitz, E. (1971) The hyaline layer: its isolation and role in echinoderm development. *Dev. Biol.*, **24**, 348–62.

Clark, W.H., Jr, Lynn, J.W., Yudin, A.I. and Persyn, H.O. (1980) Morphology of the cortical reaction in the eggs of *Penaeus aztecus*. *Biol. Bull.*, **158**, 175–86.

Clark, W.H., Jr., Yudin, A.I., Griffin, I.J. and Shigekawa, K. (1984) The control of gamete activation and fertilization in the marine Penaeidae, *Siconia ingentis*, in *Advances in Invertebrate Reproduction*, vol.3 (ed. W. Engels), Elsevier, New York, pp. 459–72.

Clarke, H.J. and Masui, Y. (1983) The induction of reversible and irreversible chromosome decondensation by protein synthesis inhibition during meiotic maturation of mouse oocytes. *Dev. Biol.*, **97**, 291–301.

Clegg, E.D. (1983) Mechanisms of mammalian sperm capacitation, in *Mechanism and Control of Animal Fertilization* (ed. J.F. Hartmann), Academic Press, New York, pp. 177–212.

Clegg, K.B. and Denny, P.C. (1974) Synthesis of rabbit globin in a cell-free protein synthesis system utilizing sea urchin egg and zygote ribosomes. *Dev. Biol.*, **37**, 263–72.

Clegg, K.B. and Pikó, L. (1982) RNA synthesis and cytoplasmic polyadenylation in the one-cell mouse embryo. *Nature*, **295**, 342–5.

Collier, J.R. (1976) Nucleic acid chemistry in the *Ilyanassa* embryo. *Am. Zool.*, **16**, 483–500.

Collins, F. (1976) A re-evaluation of the fertilizin hypothesis of sperm agglutination and the description of a novel form of sperm adhesion. *Dev. Biol.*, **49**, 381–94.

Colwin, L.H. and Colwin, A.L. (1967) Membrane fusion in relation to sperm–egg assocation, in *Fertilization*, vol. 1. (eds C.B. Metz and A. Monroy) Academic Press, New York, pp. 295–367.

Conrad, G.W. and Williams, D.C. (1974a) Polar lobe formation and cytokinesis in fertilized eggs of *Ilyanassa obsoleta*. I. Ultrastructure and effects of cytochalasin B and colchicine. *Dev. Biol.*, **36**, 363–78.

Conrad, G.W. and Williams, D.C. (1974b) Polar lobe formation and cytokinesis in fertilized eggs of *Ilyanassa obsoleta*. II. Large bleb formation caused by high concentrations of exogenous calcium ions. *Dev. Biol.*, **37**, 280–94.

Cross, N.L. and Elinson, R.P. (1980) A fast block to polyspermy in frogs mediated by changes in the membrane potential. *Dev. Biol.*, **75**, 187–98.

Daentl, D.L. and Epstein, C.J. (1971) Developmental interrelationships of uridine uptake, nucleotide formation and incorporation into RNA by early mammalian embryos. *Dev. Biol.*, **24**, 428–42.

Dale, B. and Monroy, A. (1981) How is polyspermy prevented? *Gamete Res.*, **4**, 151–69.

Dan, J.C. (1967) Acrosome reaction and lysins, in *Fertilization*, vol. 1, (eds. C.B. Metz and A. Monroy), Academic Press, New York, pp. 237–93.

Dan, K. and Nakajima, T. (1965) On the morphology of the mitotic apparatus isolated from echinoderm eggs. *Embryologia*, **3**, 187–200.

Danilchik, M.V. and Hille, M.B. (1981) Sea urchin egg and embryo ribosomes: Differences in translational activity in a cell-free system. *Dev. Biol.*, **84**, 291–8.

Das, N.K., Micou-Eastwood, J. and Alfert, M. (1975) Cytochemical studies on the protamine-type protein transition in sperm nuclei after fertilization and the early embryonic histones of *Urechis caupo*. *Dev. Biol.*, **43**, 333–9.

Davidson, E.H. (1976) *Gene Activity in Early Development*, Academic Press, New York.

Dawid, I.B. and Blackler, A.M. (1972) Maternal and cytoplasmic inheritance of mitochondrial DNA in *Xenopus*. *Dev. Biol.*, **29**, 152–61.

Decker, S.J. and Kinsey, W.H. (1983) Characterization of cortical secretory vesicles from the sea urchin egg. *Dev. Biol.*, **96**, 37–45.

Denny, P.C. and Tyler, A. (1964) Activation of protein biosynthesis in non-nucleate fragments of sea urchin eggs. *Biochem. Biophys. Res. Commun.*, **14**, 245–9.

De Petrocellis, B. and Rossi, M. (1976) Enzymes of DNA biosynthesis in developing sea urchins: Changes in ribonucleotide reductase, thymidine and thymidylate kinase activities. *Dev. Biol.*, **48**, 250–7.

De Santis, R., Jammuno, G. and Rosati, F. (1980) A study of the chorion and the follicle cells in relation to the sperm–egg interaction in the ascidian *Ciona intestinalis*. *Dev. Biol.*, **74**, 490–9.

Detering, N.K., Decker, G.L., Schmell, E.D. and Lennarz, W.L. (1977) Isolation and characterization of plasma membrane-associated cortical granules from sea urchin eggs. *J. Cell Biol.*, **75**, 899–914.

Dettlaff, T.A., Nikitina, L.A. and Stroeva, O.G. (1964) The role of the germinal vesicle in oocyte maturation in anurans as revealed by the removal and transplantation of nuclei. *J. Embryol. Exp. Morphol.*, **12**, 851–73.

Dewel, W.C. and Clark, W.H. (1974) A fine structural investigation of surface specializations and the cortical reaction in eggs of the cnidarian *Bunodosoma cavernata*. *J. Cell Biol.*, **60**, 78–91.

Dirksen, E.R. (1961) The presence of centrioles in artificially activated sea urchin eggs. *J. Biophys. Biochem. Cytol.*, **11**, 244–52.

Dohmen, M.R and Van Der Mey, J.C. (1977) Local surface differentiations at the vegetal pole of the eggs of *Nassarius reticulatus*, *Buccinum undatum* and *Crepidula fornicata* (Gastropada, Prosobranchia). *Dev. Biol.*, **61**, 104–13.

Donovan, M. and Hart, N.H. (1982) Uptake of ferritin by the mosaic egg surface of *Brachydanio*. *J. Exp. Zool.*, **223**, 299–304.

Dorée, M. and Guerrier, P. (1975) Site of action of 1-methyladenine in inducing oocyte maturation in starfish. *Exp. Cell Res.*, **91**, 296–300.

Dunbar, B.S., Munoz, M.G., Cordle, C.T. and Metz, C.B. (1976) Inhibition of fertilization *in vitro* by treatment of rabbit spermatozoa with univalent isoantibodies to rabbit sperm hyaluronidase. *J. Reprod. Fert.*, **47**, 381–84.

Ecklund, P.S. and Levine, L. (1975) Mouse sperm basic nuclear protein, electrophoretic characterization and fate after fertilization. *J. Cell Biol.*, **66**, 251–62.

Eisen, A., Kiehart, D.P., Wieland, S.J. and Reynolds, G.T. (1984) Temporal sequence and spatial distribution of early events of fertilization in single sea urchin eggs. *J. Cell Biol.*, **99**, 1647–54.

Elinson, R.P. (1980) The amphibian egg cortex in fertilization and early development, in *The Cell Surface: Mediator of Developmental Processes*. (eds. S. Subtelny and N.K. Wessels), Academic Press, New York, pp. 217–34.

Eng, L.A. and Metz, C.B. (1980) Sperm head decondensation by a high molecular weight fraction of sea urchin egg homogenates. *J. Exp. Zool.*, **212**, 159–67.

Epel, D. (1967) Protein synthesis in sea urchin eggs: A 'late' response to fertilization. *Proc. Natl. Acad. Sci., USA*, **57**, 899–906.

Epel, D. (1969) Does ADP regulate respiration following fertilization of sea urchin eggs? *Exp. Cell Res.*, **58**, 312–19.

Epel, D. (1978) Mechanisms of activation of sperm and egg during fertilization of sea urchin gametes. *Curr. Top. Dev. Biol.*, **12**, 185–246.

Epel, D. (1980) Experimental analysis of the role of intracellular calcium in the activation of the sea urchin egg at fertilization, in *The Cell Surface: Mediator of Developmental Processes.* (eds S. Subtelny and N.K. Wessels), New York: Academic pp. 169–185.

Epel, D., Steinhardt, R.A., Humphreys, T. and Mazia, D. (1974) An analysis of the partial metabolic depression of sea urchin eggs by ammonia: the existence of independent pathways. *Dev. Biol.*, **40**, 245–55.

Epel, D., Patton, C., Wallace, R.W. and Cheung, W.Y. (1981) Calmodulin activates NAD kinase of sea urchin eggs: An early event of fertilization. *Cell*, **23**, 543–9.

Eppig, J.J. and Downs, S.M. (1984) Chemical signals that regulate mammalian oocyte maturation. *Biol. Reprod.*, **30**, 1–11.

Epstein, C.J. and Daentl, D.L. (1971) Precursor pools and RNA synthesis in preimplantation mouse embryos. *Dev. Biol.*, **26**, 517–24.

Epstein, C.J., Daentl, D.L., Smith, S.A. and Kwok, L.W. (1971) Guanine metabolism in preimplantation mouse embryos. *Biol. Reprod.*, **5**, 308–13.

Ezzell, R.M. and Szego, C.M. (1979) Luteinizing hormone-accelerated redistribution of lysosome-like organelles preceding dissolution of the nuclear envelope in rat oocytes maturing *in vitro. J. Cell Biol.*, **82**, 264–77.

Fankhauser, G (1948) The organization of the amphibian egg during fertilization and cleavage. *Ann. NY Acad. Sci.*, **49**, 684–708.

Fansler, B. and Loeb, L.A. (1969) Sea urchin nuclear DNA polymerase. II Changing localization during early development. *Exp. Cell Res.*, **57**, 305–310.

Fansler, B. and Loeb, L.A. (1972) Sea urchin nuclear DNA polymerase. IV Reversible association of DNA polymerase with nuclei during the cell cycle. *Exp. Cell Res.*, **75**, 433–41.

Ferguson, J.E. and Shen, S.S. (1984) Evidence of phospholipase A2 in the sea urchin egg: Its possible involvement in the cortical granule reaction. *Gam. Res.*, **9**, 329–38.

Feuchter, F.A., Vernon, R.B. and Eddy, E.M. (1981) Analysis of the sperm surface with monoclonal antibodies: Topographically restricted antigens appearing in the epididymis. *Biol. Reprod.*, **24**, 1099–110.

Firtel, R.A. and Monroy, A. (1970) Polysomes and RNA synthesis during early development of the surf clam *Spisula solidissima. Dev. Biol.*, **21**, 87–104.

Fisher, G.W. and Rebhun, L.I. (1983) Sea urchin egg cortical granule exocytosis is followed by a burst of membrane retrieval via uptake into coated vesicles. *Dev. Biol.*, **99**, 456–72.

Florman, H. and Storey, B. (1982) Mouse gamete interactions: the zona pellucida is the site of the acrosome reaction leading to fertilization *in vitro. Dev. Biol.*, **91**, 121–30.

Florman, H.M., Bechtol, K.B. and Wassarman, P.M. (1984) Enzymatic dissection of the functions of the mouse egg's receptor for sperm. *Dev. Biol.*, **106**, 243–55.

Foerder, C.A. and Shapiro, B.M. (1977) Release of ovoperoxidase from sea urchin eggs hardens the fertilization membrane with tryosine crosslinks. *Proc. Natl. Acad. Sci., USA*, **74**, 4214–18.

Foerder, D.A., Klebanoff, S.J. and Shapiro, B.M. (1978) Hydrogen peroxide production, chemiluminescence, and the respiratory burst of fertilization: Interrelated events in early sea urchin development. *Proc. Natl. Acad. Sci., USA*, **75**, 3183–7.

Franke, W.W. (1974) Structure, biochemistry, and functions of the nuclear envelope. *Int. Rev. Cytol.*, Suppl. **4**, 72–236.

Freeman, G. and Miller, R.L. (1982) Hydrozoan eggs can only be fertilized at the site of polar body formation. *Dev. Biol.*, **94**, 142–52.

Friend, D.S. (1980) Freeze-fracture alterations in guinea pig sperm membranes preceding gamete fusion, in *Membrane–Membrane Interactions* (ed. N.B. Gilula) Raven Press, New York, pp. 153–65.

Friend, D.S., Orci, L., Perrelet, A. and Yanagimachi, R. (1977) Membrane particle changes attending the acrosome reaction in guniea pig spermatozoa. *J. Cell Biol.*, **74**, 561–77.

Frye, L.D. and Edidin, M. (1970). The rapid intermixing of cell surface antigens after formation of mouse-human heterokaryons. *J. Cell Sci.*, **7**, 319–35.

Gabel, C.A., Eddy, E.M. and Shapiro, B.M. (1979) After fertilization, sperm surface components remain as a patch in sea urchin and mouse embryos. *Cell*, **18**, 207–15.

Gaddum-Rosse, P. and Blandau, R.J. (1972) Comparative studies on the proteolysis of fixed gelatin membranes by mammalian sperm acrosomes. *Amer. J. Anat.*, **134**, 133–44.

Garbers, D.L. and Hardman, J.G. (1976) Effects of egg factors on cyclic nucleotide metabolism in sea urchin sperm. *J. Cyclic Nucleotide Res.*, **2**, 59–70.

Garbers, D.L. and Kopf, G.S. (1980) The regulation of spermatozoa by calcium and cyclic nucleotides. *Adv. Cyclic Nucleotide Res.*, **13**, 251–306.

Garbers, D.L., First, N.L. and Lardy, H.A. (1973) The stimulation of bovine epididymal sperm metabolism by cyclic nucleotide phosphodiesterase inhibitors. *Biol. Reprod.*, **8**, 589–98.

Garbers, D.L., First, N.L., Gorman, S.K. and Lardy, H.A. (1973) The effects of cyclic nucleotide phosphodiesterase inhibitors on ejaculated porcine spermatozoan metabolism. *Biol. Reprod.*, **8**, 599–606.

Gardiner, D.M. and Grey, R.D. (1983) Membrane junctions in *Xenopus* eggs: Their distribution suggests a role in calcium regulation. *J. Cell Biol.*, **96**, 1159–63.

Garner, D.L., Easton, M.P., Munson, M.E. and Doane, M.A. (1975) Immunofluorescent localization of bovine acrosin. *J. Exp. Zool.*, **191**, 127–31.

Gilkey, J.C., Jaffe, L.F., Ridgeway, E.G. and Reynolds, G.T. (1978) A free calcium wave traverses the activating egg of the medaka, *Oryzias latipes. J. Cell Biol.*, **76**, 448–66.

Girard, J., Payan, P. and Sardet, C. (1982) Changes in intracellular cations following fertilization of sea urchin eggs. Na^+/H^+ and Na^+/K^+ exchanges. *Exp. Cell Res.*, **142**, 215–21.

Giudice, G. (1973) *Developmental Biology of the Sea Urchin Embryo.* Academic Press, New York.

Glabe, C.G. and Lennarz, W.J. (1981) Isolation and partial characterization of a high molecular weight egg surface glyco-conjugate implicated in sperm adhesion. *J. Supramol. Struct.*, **15**, 387–94.

Glabe, C.G. and Vacquier, V.D. (1978) Egg surface glycoprotein receptor for sea urchin sperm bindin. *Proc. Natl. Acad. Sci., USA*, **75**, 881–5.

Golbus, M.S., Calarco, P.G. and Epstein, C.J. (1973) The effects of inhibitors of RNA synthesis (α-amanitin and actinomycin D) on preimplantation mouse embryogenesis. *J. Exp. Zool.*, **186**, 207–16.

Goldstein, J.L., Anderson, R.G.W. and Brown, M.S. (1979) Coated pits, coated vesicles, and receptor mediated endocytosis. *Nature*, **279**, 679–85.

Gordon, M., Dandekar, P.V. and Eager, P.R. (1978) Identification of phosphatases on the membranes of guinea pig sperm. *Anat. Rec.*, **191**, 123–34.

Gould-Somero, M., Holland, L. and Paul, M. (1977) Cytochalasin B inhibits sperm penetration into eggs of *Urechis caupo* (Echiura). *Dev. Biol.*, **58**, 11–22.

Graham, C.F. (1966) The regulation of DNA synthesis and mitosis in multinucleate frog eggs. *J. Cell Sci.*, **1**, 363–74.

Grainger, J.L., Winkler, M.M., Shen, S.S. and Steinhardt, R.A. (1979) Intracellular pH controls protein synthesis rate in the sea urchin egg and early embryo. *Dev. Biol.*, **68**, 396–406.

Green, G.R. and Poccia, D.L. (1985) Phosphorylation of sea urchin sperm H1 and H2B histones precedes chromatin decondensation and H1 exchange during pronuclear formation. *Dev. Biol.*, **108**, 235–45.

Green, J.D. and Summers, R.G. (1980) Ultrastructural demonstration of trypsin-like protease in acrosomes of sea urchin sperm. *Science*, **209**, 398–400.

Gross, P.R. (1964) The immediacy of genomic control during early development. *J. Exp. Zool.*, **157**, 21–38.

Gross, P.R. and Cousineau, G.H. (1964) Macromolecule synthesis and the influence of actinomycin on early development. *Exp. Cell Res.*, **33**, 368–95.

Gross, P.R., Kraemer, K. and Malkin, L.I. (1965) Base composition of RNA synthesized during cleavage of the sea urchin embryo. *Biochem. Biophys. Res. Commun.*, **18**, 569–75.

Gross, K.W., Jacobs-Lorena, M., Baglioni, C. and Gross, P.R. (1973) Cell-free translation of maternal messenger RNA from sea urchin eggs. *Proc. Natl. Acad. Sci., USA.*, **70**, 2614–18.

Gulyas, B.J. (1980) Cortical granules of mammalian eggs. *Intl. Rev. Cytol.*, **63**, 357–92.

Gulyas, B.J. and Schmell, E.D. (1980) Ovoperoxidase acitivty in ionophore treated mouse eggs. I. Electron microscopic localization. *Gam. Res.*, **3**, 267–77.

Gundersen, G.G., Gabel, C.A. and Shapiro, B.M. (1982) An intermediate state of fertilization involved in internalization of sperm components. *Dev. Biol.*, **93**, 59–72.

Gurdon, J.B. (1967) On the origin and persistence of a cytoplasmic state inducing nuclear DNA synthesis in frogs' eggs. *Proc. Natl. Acad. Sci., USA*, **58**, 545–52.

Gurdon, J.B. and Woodland, H.R. (1968) The cytoplasmic control of nuclear activity in animal development. *Biol. Rev.*, **43**, 233–67.

Gurdon, J.B., Birnstiel, M.L. and Speight, V.A. (1969) The replication of purified DNA introduced into living egg cytoplasm. *Biochim. Biophys. Acta*, **174**, 614–28.

Gwatkin, R.B.L.(1977) *Fertilization Mechanisms in Man and Mammals*. Plenum Press, New York.

Hall, H.G. (1978) Hardening of the sea urchin envelope by peroxidase catalyzed phenolic coupling of tryosines. *Cell*, **15**, 343–55.

Hamaguchi, M.S. and Hiramoto, Y. (1978) Protoplasmic movement during polar body formation in starfish oocytes. *Exp. Cell Res.*, **112**, 55–62.

Hamilton, D.W. (1977) The epididymis, in *Frontiers in Reproduction and Fertility Control* (eds R.O. Greep and M.A. Koblinsky), MIT Press, Cambridge, MA, pp. 411–26.

Hansbrough, J.R. and Garbers, D.L. (1981a) Speract. Purification and characterization of a peptide associated with eggs that activates spermatozoa. *J. Biol. Chem.*, **256**, 1447–52.

Hansbrough, J.R. and Garbers, D.L. (1981b) Sodium-dependent activation of sea urchin spematozoa by speract and monensin. *J. Biol. Chem.*, **256**, 2235–41.

Harris, P. (1979) A spiral cortical fiber system in fertilized sea urchin eggs. *Dev. Biol.*, **68**, 525–32.

Harris, P., Osborn, M. and Weber, K. (1980) A spiral array of microtubules in the fertilized sea urchin egg cortex examined by indirect immunofluorescence and electron microscopy. *Exp. Cell Res.*, **126**, 227–36.

Hartmann, J.F., Gwatkin, R.B.L. and Hutchison, C.F. (1972) Early contact interactions between mammalian gametes *in vitro*: Evidence that the vitellus influences adherence between sperm and zona pellucida. *Proc. Natl. Acad. Sci.* USA **69**, 2767–9.

Harvey, E.B. (1956) *The American Arbacia and Other Sea Urchins*, Princeton University Press, Princeton, NJ.

Hecht, N.B (1974) A DNA polymerase isolated from bovine spermatozoa. *J. Reprod. Fert.*, **41**, 345–54.

Hecht, N.B. and Williams, J.L. (1979) Nuclear and mitochondrial DNA-dependent RNA polymerases in bovine spermatozoa. *J. Reprod. Fert.*, **57**, 157–65.

Hille, M.B. (1974) Inhibitor of protein synthesis isolated from ribosomes of unfertilized eggs and embryos of sea urchins. *Nature*, **249**, 556–58.

Hille, M.B. and Albers, A.A. (1979) Efficiency of protein synthesis after fertilization of sea urchin eggs. *Nature*, **278**, 469–71.

Holland, N.D. (1979) Electron microscopic study of the cortical reaction of an ophiuroid echinoderm. *Tissue Cell*, **11**, 445–55.

Holmberg, S.R.M. and Johnson, M.H. (1979) Amino acid transport in the unfertilized and fertilized mouse egg. *J. Reprod. Fert.*, **56**, 223–31.

Hoshi, M., Numakunai, T. and Sawada, H. (1981) Evidence for participation of sperm proteinases in fertilization of the solitary ascidian, *Halocythia roretzi*: Effects of protease inhibitors. *Dev. Biol.*, **86**, 117–21.

Hoskins, D.D., Brandt, H. and Acott, T.S. (1978) Initiation of sperm motility in the mammalian epididymis. *Fed. Proc.*, **37**, 2534–42.

Hoskins, D.D., Johnson, D., Brandt, H. and Acott, T.S. (1979) Evidence for a role for a forward motility protein in the epididymal development of sperm motility, in *The Spermatozoon* (eds D.W. Fawcett and J.M. Bedford), Urban & Schwarzenberg, Baltimore, pp. 43–53.

Hosoya, H., Mabuchi, I. and Sakai, H. (1982) Actin modulating proteins in the sea urchin egg. I. Analysis of G-actin-binding proteins by DNase I-affinity chromatography and purification of a 17 000 molecular weight component. *J. Biochem*, **92**, 1853–62.

Houk, M.S. and Epel, D. (1974) Protein synthesis during hormonally induced meiotic maturation and fertilization in starfish oocytes. *Dev. Biol.*, **40**, 298–310.

Heuz, G., Marbaix, G., Hubert, E., Leclercq, M., Nudel, U., Soreq, H., Salomon, R., Leblue, B., Revel, M. and Littauer, U.A. (1974) Role of the polyadenylate segment in the translation of globin messenger RNA in *Xenopus* oocytes. *Proc. Nat. Acad. Sci., USA*, **71**, 3143–6.

Howlett (1986) A set of proteins showing cell cycle dependent modification in the early mouse embryo. *Cell*, **40**, 387–96.

Humphreys, T. (1969) Efficiency of translation of messenger RNA before and after fertilization of sea urchin eggs. *Dev. Biol.*, **20**, 435–58.

Humphreys, T. (1971) Measurements of messenger RNA entering polysomes upon fertilization of sea urchin eggs. *Dev. Biol.*, **26**, 201–08.

Humphreys, W.J. (1967) The fine structure of cortical granules in eggs and gastrulae of *Mytilus edulis. J. Ultrastruct. Res.*, **17**, 314–26.

Hunter, R.H.F. (1967) Polyspermic fertilization in pigs during the luteal phase of the estrous cycle. *J. Exp. Zool.*, **165**, 451–60.

Hunter, R.H.F. (1976) Sperm–egg interactions in the pig: Monospermy, extensive polyspermy, and the formation of chromatin aggregates. *J. Anat.*, **122**, 43–59.

Hylander, B.L. and Summers, R.G. (1981) The effect of local anesthetics and ammonia on cortical granule-plasma membrane attachment in the sea urchin egg. *Dev. Biol.*, **86**, 1–11.

Hylander, B.L. and Summers, R.G. (1982) Observations on the role of the cortical reaction in surface changes at fertilization. *Cell Diff.*, **11**, 267–70.

Hylander, B.E., Anstrom, J. and Summers, R.C. (1981) Premature sperm incorporated into the primary oocyte of the polychaete *Pectinaris*: Male pronuclear formation and oocyte maturation. *Dev. Biol.*, **82**, 382–87.

Hyne, R.V. and Garbers, D.L. (1979) Calcium-dependent increase in adenosine 3′5′-monophosphate and induction of the acrosome reaction in guinea pig spermatozoa. *Proc. Nat. Acad. Sci. USA* **76**, 5699–703.

Imai, H., Niwa, K. and Iritani, A. (1977) Penetration *in vitro* of zona-free hamster eggs by ejaculated boar spermatozoa. *J. Reprod.*, **51**, 495–7.

Iwamatsu, T. and Chang, M.C. (1972) Sperm penetration *in vitro* of mouse oocyte at various times during maturation. *J. Reprod. Fert.*, **31**, 237–47.

Jaffe, L.A. (1976) Fast block to polyspermy in sea urchin eggs is electrically mediated. *Nature*, **261**, 68–71.

Jaffe, L.A., Hagiwara, S. and Kado, R.T. (1978) The time course of cortical vesicle fusion in sea urchin eggs observed as membrane capacitance changes. *Dev. Biol.*, **67**, 243–8.

Jaffe, L.A., Sharp, A.P. and Wolf, D.P. (1983) Absence of an electrical poly-spermy block in the mouse. *Dev. Biol.*, **96**, 317–23.

Jaffe, L.F. (1983) Sources of calcium in egg activation. A review and hypothesis. *Dev. Biol.*, **99**, 265–76.

Jeffery, W.R. (1984) Pattern formation by ooplasmic segregation in ascidian eggs. *Biol. Bull*, **166**, 277–98.

Jenkins, N.A., Kaumeyer, J.F., Young, E.M. and Raff, R.A. (1978) A test for masked message: The template activity of messenger ribonucleoprotein particles isolated from sea urchin eggs. *Dev. Biol.*, **63**, 279–98.

Johnson, C.H. and Epel, D. (1982) Starfish oocyte maturation and fertilization: Intracellular pH is not involved in activation. *Dev. Biol.*, **92**, 461–9.

Johnson, J.D., Epel, D. and Paul, M. (1976) Intracellular pH and activation of sea urchin eggs after fertilization. *Nature*, **262**, 661–4.

Johnson, R.T. and Rao, P.N. (1971) Nucleo-cytoplasmic interactions in the achievement of nuclear synchrony in DNA synthesis and mitosis in multi-nucleate cells. *Biol. Rev.*, **46**, 97–155.

Johnston, R.N. and Paul, M. (1977) Calcium influx following fertilization of *Urechis caupo* eggs. *Dev. Biol.*, **57**, 364–74.

Jones, H.P., Bradford, M.M., McRorie, R.A. and Cormier, M.H. (1978) High levels of a calcium-dependent modulator protein in spermatozoa and its similarity to brain modulator protein. *Biochem. Biophys. Res. Comm.*, **82**, 1264–72.

Kanatani, H., Shirai, H., Nakanishi, K. and Kurokawa, T. (1969) Isolation and identification of meiosis inducing substance in starfish *Asterias amurensis*. *Nature*, **221**, 273–4.

Kane, R.E. (1973) Hyaline release during normal sea urchin development and its replacement after removal at fertilization. *Exp. Cell Res.*, **81**, 301–11.

Karp, G.C. (1973) Autoradiographic patterns of [^{3}H]-uridine incorporation during the development of the mollusc, *Acmaea scutum. J. Embryol. Exp. Morphol.*, **29**, 15–25.

Katagiri, C. (1974) A high frequency of fertilization in premature and mature coelomic toad eggs after enzymic removal of vitelline membrane. *J. Embryol. Exp. Morph.*, **31**, 573–87.

Katagiri, C. and Moriya, M. (1976) Spermatozoon response to the toad egg matured after removal of germinal vesicle. *Dev. Biol.*, **50**, 235–41.

Kidder, G.M. (1976) RNA synthesis and the ribosomal cistrons in early molluscan development. *Am. Zool.*, **16**, 501–20.

Kinsey, W.H., Rubin, J.A. and Lennarz, W.J. (1980) Studies on the specificity of sperm binding in echinoderm fertilization. *Dev. Biol.*, **74**, 245–50.

Kirschner, M., Gerhart, J.C., Hara, K. and Ubbels, G.A. (1980) Initiation of the cell cycle and establishment of bilateral symmetry in *Xenopus* eggs, in *The Cell surface: Mediator of Developmental Processes* (eds S. Subtelny and N.K. Wessels), Academic Press, New York, pp. 187–215.

Kishimoto, T., Hirai, S. and Kanatani, H. (1981) Role of germinal vesicle material in producing maturation-promoting factor in starfish oocyte. *Dev. Biol.*, **81**, 177–81.

Koehler, J.K. (1978) The mammalian sperm surface: Studies with specific labeling techniques. *Int. Rev. Cytol.*, **54**, 73–108.

Koehler, J.K. and Gaddum-Rosse, P. (1975) Media induced alterations of the membrane associated particles of the guinea pig sperm tail. *J. Ultrastruct. Res.*, **51**, 106–18.

Kopf, G.S. and Garbers, D.L. (1980) Calcium and fucose-sulfate rich polymer regulate sperm cyclic nucleotide metabolism and the acrosome reaction. *Biol. Reprod.*, **22**, 1118–26.

Kunkle, M., Longo, F.J. and Magun, B.E. (1978) Nuclear protein changes in the maternally and paternally derived chromatin at fertilization. *J. Exp. Zool.*, **203**, 371–80.

Kunkle, M., Magun, B.E. and Longo, F.J. (1978) Analysis of isolated sea urchin nuclei incubated in egg cytosol. *J. Exp. Zool.*, **203**, 381–90.

Lambert, C.C. and Lambert, G. (1981) The ascidian sperm reaction: Ca^{2+} uptake in relation to H^+ efflux. *Dev. Biol.*, **88**, 312–17.

Laskey, R.A. and Gurdon, J.B. (1973) Induction of polyoma DNA synthesis by injection into frog egg cytoplasm. *Eur. J. Biochem.*, **37**, 467–71.

Lee, H.C., Johnson, C. and Epel, D. (1983) Changes in internal pH associated with initiation of motility and acrosome reaction of sea urchin sperm. *Dev. Biol.*, **95**, 31–45.

Lee, S.C. and Steinhardt, R.A. (1981) pH changes associated with meiotic maturation in oocytes of *Xenopus laevis*. *Dev. Biol.*, **85**, 358–69.

Levine, A.E. and Walsh, K.A. (1978) Involvement of an acrosin-like enzyme in the acrosome reaction of sea urchin sperm. *Dev. Biol.*, **72**, 126–37.

Lillie, F.R. (1919) *Problems of Fertilization.*, University of Chicago Press, Chicago.

Loeb, L.A. and Fansler, B. (1970) Sea urchin DNA polymerase. III. Intracellular migration of DNA polymerase in early developing sea urchin embryos. *Biochim. Biophys. Acta*, **217**, 50–5.

Loeb, L.A., Fansler, B., Williams, R. and Mazia, D. (1969) Sea urchin DNA polymerase. I. Localization in nuclei during rapid DNA synthesis. *Exp. Cell Res.*, **57**, 298–304.

Lohka, M.J. and Masui, Y. (1983a) Formation *in vitro* of sperm pronuclei and mitotic chromosomes induced by amphibian ooplasmic components. *Science*, **220**, 719–21.

Lohka, M.J. and Masui, Y. (1983b) The germinal vesicle material required for sperm pronuclear formation is located in the soluble fraction of egg cytoplasm. *Exp. Cell Res.*, **148**, 481–91.

Lohka, M.J. and Masui, Y. (1984a) Effects of Ca^{2+} ions on the formation of metaphase chromosomes and sperm pronuclei in cell-free preparations from unactivated *Rana pipiens* eggs. *Dev. Biol.*, **103**, 434–42.

Lohka, M.J. and Masui, Y. (1984b) Roles of cytosol and cytoplasmic particles in nuclear envelope assembly and sperm pronuclear formation in cell-free preparations from amphibian eggs. *J. Cell Biol.*, **98**, 1222–30.

Longo, F.J. (1972) The effects of cytochalasin B on the events of fertilization in the surfclam, *Spisula solidissima*. I. Polar body formation. *J. Exp. Zool.*, **182**, 321–44.

Longo, F.J. (1973) Fertilization: A comparative ultrastructural review. *Biol. Reprod.*, **9**, 149–215.

Longo, F.J. (1976a) Ultrastructural aspects of fertilization in spiralian eggs. *Amer. Zool.*, **16**, 375–94.

Longo, F.J. (1976b) Derivation of the membrane comprising the male pronuclear envelope in inseminated sea urchin eggs. *Dev. Biol.*, **49**, 347–68.

Longo, F.J. (1976c) Sperm aster in rabbit zygotes: Its structure and function. *J. Cell Biol.*, **69**, 539–47.

Longo, F.J. (1977) An ultrastructural study of cross-fertilization (*Arbacia* ♀ × *Mytilus* ♂). *J. Cell Biol.*, **73**, 14–26.

Longo, F.J. (1978a) Effects of cytochalasin B on sperm–egg interactions. *Dev. Biol.*, **67**, 249–65.

Longo, F.J. (1978b) Insemination of immature sea urchin (*Arbacia punctulata*) eggs. *Dev. Biol.*, **62**, 271–91.

Longo, F.J. (1980) Organization of microfilaments in sea urchin *Arbacia punctulata* eggs at fertilization: Effects of cytochalasin B. *Dev. Biol.*, **74**, 422–33.

Longo, F.J. (1981a) Morphological features of the surface of the sea urchin (*Arbacia punctulata*) egg: Oolemma-cortical granule association. *Dev. Biol.*, **84**, 173–82.

Longo, F.J. (1981b) Regulation of pronuclear development, in *Bioregulators of Reproduction* (eds G. Jagiello and C. Vogel), Academic Press, New York, pp. 529–57.

Longo, F.J. (1982) Integration of sperm and egg plasma membrane components at fertilization. *Dev. Biol.*, **89**, 409–16.

Longo, F.J. (1983) Meiotic maturation and fertilization, in *Biology of Mollusca*, vol. 3, (eds K.M. Wilbur, N.H. Verdonk, J.A.M. Van den Biggelaar and A.S. Tompa), Academic Press, New York, pp. 49–89.

Longo, F.J. (1984) Transformations of sperm nuclei incorporated into sea urchin (*Arbacia punctulata*) embryos at different stages of the cell cycle. *Dev. Biol.*, **103**, 168–81.

Longo, F.J. and Anderson, C. (1968) The fine structure of pronuclear development and fusion in the sea urchin, *Arbacia punctulata*. *J. Cell Biol.*, **39**, 339–68.

Longo, F.J. and Anderson, E. (1969a) Cytological aspects of fertilization in the lamellibranch, *Mytilus edulis*. I. Polar body formation and development of the female pronucleus. *J. Exp. Zool.*, **172**, 69–96.

Longo, F.J. and Anderson, E. (1969b) Cytological events leading to the formation of the two-cell stage in the rabbit: Association of the maternally and paternally derived genomes. *J. Ultrastruct. Res.*, **29**, 86–118.

Longo, F.J. and Anderson, E. (1970a) An ultrastructural analysis of fertilization in the surf clam, *Spisula solidissima*. I. Polar body formation and development of the female pronucleus. *J. Ultrastruct. Res.*, **33**, 495–514.

Longo, F.J. and Anderson, E. (1970b) An ultrastructural analysis of fertilization in the surf clam, *Spisula solidissima*. II. Development of the male pronucleus and the association of the maternally and paternally derived chromosomes. *J. Ultrastruct. Res.*, **33**, 515–27.

Longo, F.J. and Anderson, E. (1970c) A cytological study of the relation of the cortical reaction to subsequent events of fertilization in urethane-treated eggs of the sea urchin, *Arbacia punctulata*. *J. Cell Biol.*, **47**, 646–65.

Longo, F.J. and Anderson, E. (1970d) The effects of nicotine on fertilization in the sea urchin, *Arbacia punctulata*. *J. Cell Biol.*, **46**, 308–25.

Longo, F.J. and Anderson, E. (1974) Gametogenesis, in *Concepts of Develop-*

ment (eds J. Lash and J.R. Whittaker), Sinauer Assoc., Stanford, CN. pp. 3–47.

Longo, F.J. and Chen, D.Y. (1985) Development of cortical polarity in mouse eggs: Involvement of the meiotic apparatus. *Dev. Biol.*, **107**, 382–94.

Longo, F.J. and Kunkle, M. (1977) Synthesis of RNA by male pronuclei of fertilized sea urchin eggs. *J. Exp. Zool.*, **201**, 431–8.

Longo, F.J. and Kunkle, M. (1978) Transformation of sperm nuclei upon insemination. *Curr. Top. Dev. Biol.*, **12**, 149–84.

Longo, F. J., Lynn, J.W., McCulloh, D.H. and Chambers, E.L. (1986) Correlative ultrastructural and electrophysiological studies of sperm-egg interactions of the sea urchin, *Lytechinus variegatus. Dev. Biol.*, **118**, 155–66.

Longo, F.J. and Plunkett, W. (1973) The onset of DNA synthesis and its relation to morphogenetic events of the pronuclei in activated eggs of the sea urchin, *Arbacia punctulata. Dev. Biol.*, **30**, 56–67.

Lopata, A., Sathananthan, A.H., McBain, J.C., Johnston, W.I.H. and Speirs, A.L. (1980) The ultrastructure of the preovulatory human egg fertilized *in vitro. Fertil. Steril.*, **33**, 12–20.

Lopo, A.C. (1983) Sperm–egg interactions in invertebrates, in *Mechanism and Control of Animal Fertilization.* (ed. J.F. Hartmann), Academic Press, New York, pp. 269–324.

Lopo, A.C. and Vacquier, V.D. (1980) Antibody to a sperm surface glycoprotein inhibits the egg jelly-induced acrosome reaction of sea urchin sperm. *Dev. Biol.*, **79**, 325–33.

Luchtel, D.L. (1976) An ultrastructural study of the egg and early cleavage stages of *Lymnaea stagnalis,* a pulmonate mollusc. *Am. Zool.*, **16**, 405–19.

Luthardt, F.W. and Donahue, R.P. (1973) Pronuclear DNA synthesis in mouse eggs: An autoradiographic study. *Exp. Cell Res.*, **82**, 143–51.

Luttmer, S. and Longo, F.J. (1985) Ultrastructural and morphometric observations of cortical endoplasmic reticulum in *Arbacia, Spisula* and mouse eggs. *Dev. Growth Differ.*, **27**, 349–59.

Mabuchi, I. (1973) ATPase in the cortical layer of sea urchin egg, its properties and interaction with cortex protein. *Biochim. Biophys. Acta*, **297**, 317–32.

Mabuchi, I. (1981) Purification from starfish eggs of a protein that depolymerizes actin. *J. Biochem.*, **89**, 1341–4.

Mabuchi, I., Hamaguchi, Y., Kobayashi, T., Hosoya, H., Tsukita, S. and Tsukita, S. (1985) Alpha-actinin from sea urchin eggs: Biochemical properties, interaction with actin, and distribution in the cell during fertilization and cleavage. *J. Cell Biol.*, **100**, 375–583.

Maller, J.L. and Krebs, E.G. (1977) Progesterone-stimulated meiotic cell division in *Xenopus* oocytes: Induction by regulatory subunit and inhibition by catalytic subunit of adenosine 3′:5′-monophosphate-dependent protein kinase. *J. Biol. Chem.*, **252**, 1712–18.

Maller, J.L. and Krebs, E.G. (1980) Regulation of oocyte maturation. *Curr. Top. Cell Reg.*, **16**, 271–311.

Maller, J., Wu, M. and Gerhart, J.C. (1977) Changes in protein phosphorylation accompanying maturation of *Xenopus laevis* oocytes. *Dev. Biol.*, **58**, 295–312.

Manes, C. (1973) The participation of the embryonic genome during early cleavage in the rabbit. *Dev. Biol.*, **32**, 453–9.

Mar, H. (1980) Radial cortical fibers and pronuclear migration in fertilized and artificially activated eggs of *Lytechinus pictus*. *Dev. Biol.*, **78**, 1–13.

Maro, B., Johnson, M.H., Pickering, S.J. and Flach, G. (1984) Changes in actin distribution during fertilization of the mouse egg. *J. Embryol. Exp. Morphol.*, **81**, 211–37.

Masui, Y. and Clarke, H. (1979) Oocyte maturation. *Int. Rev. Cytol.*, **57**, 185–282.

Masui, Y. and Markert, C.L. (1971) Cytoplasmic control of nuclear behavior during meiotic maturation of frog oocytes. *J. Exp. Zool.*, **177**, 129–46.

Mazia, D. (1937) The release of calcium in *Arbacia* eggs on fertilization. *J. Cell Comp. Physiol.*, **10**, 291–304.

Mazia, D. (1961) Mitosis and the physiology of cell division, in *The Cell*, vol. III (eds J. Brachet and A.E. Mirsky), Academic Press, New York, pp. 77–412.

Mazia, D. and Ruby, A. (1974) DNA synthesis turned on in unfertilized sea urchin eggs by treatment with NH_4OH. *Exp. Cell Res.*, **85**, 164–72.

McCulloh, D.H., Rexroad, C.E. and Levitan, H. (1983) Insemination of rabbit eggs is associated with slow depolarization and repetitive diphasic membrane potentials. *Dev. Biol.*, **95**, 372–7.

McIntosh, J.R. (1983) The centrosome as an organizer of the cytoskeleton, in *Spatial Organization of Eukaryotic Cells* (ed. J.R. McIntosh) Alan R. Liss, New York, pp. 115–42.

McLean, K. (1976) Some aspects of RNA synthesis in oyster development. *Am. Zool.*, **16**, 521–8.

McLean, K.W. and Whiteley, A.H. (1974) RNA synthesis during the early development of the Pacific oyster, *Crassostrea gigas*. *Exp. Cell Res.*, **87**, 132–8.

Meizel, S. and Mukerji, S.K. (1975) Proacrosin from rabbit epididymal spermatozoa: Partial purification and initial biochemical characterization. *Biol. Reprod.*, **13**, 83–93.

Merriam, R.W. (1969) Movement of cytoplasmic proteins into nuclei induced to enlarge and initiate DNA or RNA synthesis. *J. Cell Sci.*, **5**, 333–49.

Metz, C.B. (1967) Gamete surface components and their role in fertilization, in *Fertilization*, vol. 1. (eds C.B. Metz and A. Monroy), Academic Press, New York, pp. 163–236.

Metz, C.B. and Monroy, A. (eds.) (1985) *Fertilization*, vols 1–3, Academic Press, New York.

Meyerhof, P.G. and Masui, Y. (1977) Ca and Mg control of cytostatic factors from *Rana pipiens* oocytes which cause metaphase and cleavage arrest. *Dev. Biol.*, **61**, 214–29.

Miller, R.L. (1977) Distribution of sperm chemotaxis in the animal kingdom. *Adv. Invertebr. Reprod.*, **1**, 99–119.

Millonig, G. (1969) Fine structure analysis of the cortical reaction in the sea urchin egg after normal fertilization and after electric inducation. *J. Submicr. Cytol.*, **1**, 69–84.

Mintz, B. (1964) Synthetic processes and early development in the mammalian egg. *J. Exp. Zool.*, **157**, 85–100.

Mirkes, P.E. (1970) Protein synthesis before and after fertilization in the egg of *Ilyanassa obsoleta*. *Exp. Cell Res.*, **60**, 115–18.

Monroy, A. (1957) Adenosinetriphosphatase in the mitochondria of unfertilized and newly fertilized sea-urchin eggs. *J. Cell. Comp. Physiol.*, **50**, 73–81.

Monroy, A. (1965) *Chemistry and Physiology of Fertilization*, Holt, Rinehart & Winston, New York.

Moore, H.D.M. and Bedford, J.M. (1983) The interaction of mammalian gametes in the female, in *Mechanism and Control of Animal Fertilization*, (ed. J.F. Hartmann), Academic Press, New York, pp. 453–97.

Moreau, M., Guerrier, P., Doree, M. and Ashley, C.C. (1978) Hormone-induced release of intracellular Ca^{2+} triggers meiosis in starfish oocytes. *Nature*, **272**, 251–3.

Moriya, M. and Katagiri, C. (1976) Microinjection of toad sperm into oocytes undergoing maturation division. *Dev. Growth Differ.*, **18**, 349–56.

Morton, D.B. (1975) Acrosomal enzymes: Immunological localization of acrosin and hyaluronidase in ram spermatozoa. *J. Reprod. Fert.*, **45**, 375–8.

Motlik, J. and Fulka, J. (1974) Fertilization of pig follicular oocytes cultivated *in vitro*. *J. Reprod. Fert.*, **36**, 235–7.

Motlik, J., Kopecny, V., Pivko, J. and Fulka, J. (1980) Distribution of proteins labeled during meiotic maturation in rabbit and pig eggs at fertilization. *J. Reprod. Fert.*, **58**, 415–19.

Moy, G.W., Kopf, G.S., Gache, C. and Vacquier V.D. (1983) Calcium-mediated release of glucanase activity from cortical granules of sea urchin eggs. *Dev. Biol.*, **100**, 267–74.

Myles, D.G. and Primakoff, P. (1984) Localized surface antigens of guinea pig sperm migrate to new regions prior to fertilization. *J. Cell. Biol.*, **99**, 1634–41.

Newrock, K.M., Alfageme, C.R., Nardi, R.V. and Cohen, L.H. (1977) Histone changes during chromatin remodeling in embryogenesis. *Cold Spring Harbor Symp. Quant. Biol.*, **42**, 421–31.

Nicolson, G.L. and Yanagimachi, R. (1979) Cell surface changes associated with the epididymal maturation of mammalian spermatozoa, in *The Spermatozoon* (eds D.W. Fawcett and J.M. Bedford), Urban & Schwarzenberg, Baltimore, pp. 187–94.

Nicosia, S.V., Wolf, D.P. and Inoue, M. (1977) Cortical granule distribution and cell surface characteristics in mouse eggs. *Dev. Biol.*, **57**, 56–74.

O'Connor, C.M., Robinson, K.R. and Smith, L.D. (1977) Calcium, potassium, and sodium exchange by full-grown and maturing *Xenopus laevis* oocytes. *Dev. Biol.*, **61**, 28–40.

O'Dell, D.S., Ortolani, G. and Monroy, A. (1973) Increased binding of radioactive concanavalin A during maturation of *Ascidia* eggs. *Exp. Cell Res.*, **83**, 408–11.

Olson, G.E. and Danzo, B.J. (1981) Surface changes in rat spermatozoa during epididymal transit. *Biol. Reprod.*, **24**, 431–43.

Oprescu, S. and Thibault, C. (1965) Duplication de l'ADN dans les oeufs de lapine aprés la fécondation. *Ann. Biol. Anim. Biochim. Biophys.*, **5**, 151–6.

Ortolani, G., O'Dell, D.S., Mansueto, C. and Monroy, A. (1975) Surface changes and onset of DNA replication in the *Ascidia* egg. *Exp. Cell Res.*, **96**, 122–8.

Otto, J.J. and Schroeder, T.E. (1984) Microtubule arrays in the cortex and near the germinal vesicle of immature starfish oocytes. *Dev. Biol.*, **101**, 274–81.

Otto, J.J., Kane, R.E. and Bryan, J. (1980) Redistribution of actin and fascin in sea urchin eggs after fertilization. *Cell Motil.*, **1**, 31–40.

Paiement, J., Beaufay, H. and Godelaine, D. (1980) Coalescence of microsomal vesicles from rat liver: A phenomenon occurring in parallel with enhancement of the glycosylation activity during incubation of stripped rough microsomes with GTP. *J. Cell Biol.*, **86**,. 29–37.

Pasteels, J.J. (1965) Etude au microscope électronique de la réaction corticale. I. La réaction corticale de fécondation chez *Paracentrotus* et sa chronologie. II. La réaction corticale de l'oeuf vierge de *Sabellaria alveolata*. *J. Embryol. Exp. Morph.*, **13**, 327–40.

Paul, M. (1975) Release of acid and changes in light-scattering properties following fertilization of *Urechis caupo* eggs. *Dev. Biol.*, **43**, 299–312.

Paul, M. and Johnston, R.N. (1978) Uptake of Ca^{2+} is one of the earliest responses to fertilization of sea urchin eggs. *J. Exp. Zool.*, **203**, 143–9.

Payan, P., Girard, J.-P. and Ciapa, B. (1983) Mechanisms regulating intracellular pH in sea urchin eggs. *Dev. Biol.*, **100**, 29–38.

Perreault, S., Zaneveld, L.J.D. and Rogers, B.J. (1979) Inhibition of fertilization in the hamster by sodium aurothiomalate, a hyaluronidase inhibitor. *J. Reprod. Fert.*, **60**, 461–7.

Phillips, S.G., Phillips, D.M., Dev, V.G. et al. (1976) Spontaneous cell hybridization of somatic cells present in sperm suspensions. *Exp. Cell Res.*, **98**, 429–43.

Picheral, B. (1977) La fécodation chez le Triton *pleurodele*. II. La pénétration des spermatozoides et al réaction locale de l'oeuf. *J. Ultrastruct. Res.*, **60**, 181–202.

Poccia, D., Krystal, G., Nishioka, D. and Salik, J. (1978) Control of sperm chromatin structure by egg cytoplasm in the sea urchin, in *ICN–UCLA Symposium on Molecular and Cellular Biology: Cell Reproduction, XII* (eds E.R. Dirksen, D.M. Prescott and C.F. Fox), Academic Press, New York, pp. 197–206.

Poccia, D., Salik, J. and Krystal, G. (1981) Transitions in histone variants of the male pronucleus following fertilization and evidence for a maternal store of cleavage-stage histones in sea urchin egg. *Dev. Biol.*, **82**, 287–96.

Pollard, T.D. and Craig, S.W. (1982) Mechanisms of actin polymerization. *Trends Biochem. Sci.*, **7**, 55–8.

Raff, R.A. (1980) Masked messenger RNA and the regulation of protein synthesis in eggs and embryos, in *Cell Biology: A Comprehensive Treatise*, Vol. 4, (eds D.M. Prescott and L. Goldstein), Academic Press, New York, pp. 107–36.

Raff, R.A., Greenhouse, G., Gross, K.W. and Gross, P.R. (1971) Synthesis and storage of microtubule proteins by sea urchin embryos. *J. Cell Biol.* 50, 516–27.

Rappaport, R. (1971) Cytokinesis in animal cells. *Int. Rev. Cytol.*, **31**, 169–213.

Raven, C.P. (1966) *Morphogenesis: The Analysis of Molluscan Development*, Pergamon Press, Oxford.

Raven, C.P. (1972) Chemical embryology of mollusca, in *Chemical Zoology*

Vol. 7, (eds M. Florkin and B.T. Scheer), Academic Press, New York, pp. 155–85.

Rodman, T.C., Pruslin, F.H., Hoffmann, H.P. and Allfrey, V.G. (1981) Turnover of basic chromosomal proteins in fertilized eggs: A cytoimmunochemical study of events *in vivo. J. Cell Biol.,* **90**, 351–61.

Rosati, F., Monroy, A. and De Prisco, P. (1977) Fine structural study of fertilization in the ascidian *Ciona intestinalis. J. Ultrastruct. Res.,* **58**, 261–70.

Rosenthal, E.T., Hunt, T. and Ruderman, J.V. (1980) Selective translation of mRNA controls the pattern of protein synthesis during early development of the surf clam, *Spisula solidissima. Cell,* **20**, 487–94.

Ruderman, J.V. and Gross, P.R. (1974) Histones and histone synthesis in sea urchin development. *Dev. Biol.,* **36**, 286–98.

Ruderman, J.V. and Pardue, M.L. (1977) Cell-free translation analysis of messenger RNA in echinoderm and amphibian development. *Dev. Biol.,* **60**, 48–68.

Russell, L., Peterson, R.N. and Freund, M. (1979) Direct evidence for formation of hybrid vesicles by fusion of plasma and outer acrosomal membranes during the acrosome reaction in boar spermatozoa. *J. Exp. Zool.,* **208**, 41–56.

Saling, P.M., Eckberg, W.R. and Metz, C.B. (1982) Mechanism of univalent anti-sperm antibody inhibition of fertilization in the sea urchin, *Arbacia punctulata. J. Exp. Zool.,* **221**, 93–9.

Sano, K. and Kanatani, H. (1980) External calcium ions are requisite for fertilization of sea urchin eggs by spermatozoa with reacted acrosomes. *Dev. Biol.,* **78**, 242–6.

Sardet, C. (1984) The ultrastructure of the sea urchin egg cortex isolated before and after fertilization. *Dev. Biol.,* **105**, 196–210.

Sardet, C., Carré, D., Cosson, M.P. et al. (1982) Some aspects of fertilization in marine invertebrates. *Prog. Clin. Biol. Res.,* **91**, 185–210.

Sargent, T.D. and Raff, R.A. (1976) Protein synthesis and messenger RNA stability in activated, enucleate sea urchin eggs are not affected by actinomycin D. *Dev. Biol.,* **48**, 327–35.

Sasaki, H. (1984) Modulation of calcium sensitivity by a specific cortical protein during sea urchin egg cortical vesicle exocytosis. *Dev. Biol.,* **101**, 125–35.

Schackmann, R.W. and Shapiro, B.M. (1981) A partial sequence of ionic changes associated with the acrosome reaction of *Strongylocentrotus purpuratus. Dev. Biol.,* **81**, 145–54.

Schackmann, R.W., Eddy, E.M. and Shapiro, B.M. (1978) The acrosome reaction of *Strongylocentrotus purpuratus* sperm. Ion requirements and movements. *Dev. Biol.,* **65**, 483–95.

Schatten, G. (1984) The supramolecular organization of the cytoskeleton during fertilization, in *Subcellular Biochemistry, Vol. 1* (ed. D.B. Roodyn), Plenum Press, New York, pp. 359–453.

Schmell, E.D. and Gulyas, B.J. (1980) Ovoperoxidase activity in ionophore treated mouse eggs. II. Evidence for the enzyme's role in hardening the zona pellucida. *Gam. Res.* **3**, 279–90.

Schmell, E.D., Earles, B.J., Breaux, C. and Lennarz, W.J. (1977) Identification of a sperm receptor on the surface of the eggs of the sea urchin *Arbacia punctulata*. *J. Cell. Biol.*, **72**, 35–46.

Schmell, E.D., Gulyas, B.J. and Hedrick, J.L. (1983) Egg surface changes during fertilization and the molecular mechanism of the block to polyspermy, in *Mechanism and Control of Animal Fertilization* (ed. J.F. Hartmann), Academic Press, New York, pp. 365–413.

Schmidt, T., Patton, C. and Epel, D. (1982) Is there a role for the Ca^{2+} influx during fertilization of the sea urchin egg? *Dev. Biol.* **90**, 284–90.

Schneider, E.G. (1985) Activation of Na^+-dependent transport at fertilization in the sea urchin: Requirements of both an early event associated with exocytosis and a later event involving increased energy metabolism. *Dev. Biol.*, **108**, 152–63.

Schroeder, T.E. (1975) Dynamics of the contractile ring, in *Molecules and Cell Movement* (eds S. Inoué and R.E. Stephens), Raven Press, New York, pp. 305–34.

Schroeder, T.E. (1979) Surface area change at fertilization: Resorption of the mosaic membrane. *Dev. Biol.*, **70**, 306–26.

Schroeder, T.E. (1980) Expression of the prefertilization polar axis in sea urchin eggs. *Dev. Biol.*, **79**, 428–43.

Schroeder, T.E. and Otto, J.J. (1984) Cyclic assembly-disassembly of cortical microtubules during maturation and early development of starfish oocytes. *Dev. Biol.*, **103**, 493–503.

Schuel, H. (1978) Secretory functions of egg cortical granules in fertilization and development. A critical review. *Gam. Res.*, **1**, 299–382.

Schuel, H., Kelly, J.W., Berger, E.R. and Wilson, W.L. (1974) Sulfated acid mucopolysaccharides in the cortical granules of eggs. Effects of quaternary ammonium salts on fertilization. *Exp. Cell Res.*, **88**, 24–30.

Schuetz, A.W. (1975) Induction of nuclear breakdown and meiosis in *Spisula solidissima* oocytes by calcium ionophore. *J. Exp. Zool.*, **191**, 433–40.

SeGall, G.K. and Lennarz, W.J. (1979) Chemical characterization of the component of the jelly coat from sea urchin eggs responsible for induction of the acrosome reaction. *Dev. Biol.*, **71**, 33–48.

Shalgi, R. and Phillips, D.M. (1980a) Mechanics of sperm entry in cyclic hamster. *J. Ultrastruct. Res.*, **71**, 154–61.

Shalgi, R. and Phillips, D.M. (1980b) Mechanics of *in vitro* fertilization in the hamster. *Biol. Reprod.*, **23**, 433–44.

Shapiro, B.M. (1975) Limited proteolysis of some egg components is an early event following fertilization of the sea urchin, *Strongylocentrotus purpuratus*. *Dev. Biol.*, **46**, 88–102.

Shapiro, B.M. (1981) Awakening of the invertebrate egg at fertilization, in *Fertilization and Embryonic Development In Vitro* (eds L. Mastroianni and J.D. Biggers), Plenum Press, New York, pp. 233–55.

Shapiro, B.M. and Eddy, E.M. (1980) When sperm meets egg: Biochemical mechanisms of gamete interaction. *Intl. Rev. Cytol.*, **66**, 257–302.

Shen, S.S. (1983) Membrane properties and intracellular ion activities of marine invertebrate eggs and their changes during activation, in *Mechanism and Control of Animal Fertilization* (ed. J.F. Hartmann), Academic Press, New York, pp. 213–67.

Shen, S.S. and Steinhardt, R.A. (1978) Direct measurement of the intracellular pH during metabolic derepression of the sea urchin egg. *Nature*, **272**, 253–4.

Sherman, M.I. (1979) Developmental biochemistry of pre-implantation mammalian embryos. *Ann. Rev. Biochem.*, **48**, 443–70.

Shih, R.J., O'Connor, C.M., Keem, K. and Smith, L.D. (1978) Kinetic analysis of amino acid pools and protein synthesis in amphibian oocytes and embryos. *Dev. Biol.*, **66**, 172–82.

Shimizu, T. (1979) Surface contractile activity of the *Tubifex* egg: Its relationship to the meiotic apparatus function. *J. Exp. Zool.*, **208**, 361–78.

Shimizu, T. (1981a) Cortical differentiation of the animal pole during maturation division in fertilized eggs of *Tubifex* (Annelida, Oligochaeta). I. Meiotic apparatus formation. *Dev. Biol.*, **85**, 65–76.

Shimizu, T. (1981b) Cortical differentiation of the animal pole during maturation division in fertilized eggs of *Tubifex* (Annelida, Oligochaeta). II. Polar body formation. *Dev. Biol.*, **85**, 77–88.

Showman, R.M., Wells, D.E., Anstrom, J et al (1982) Message-specific sequestration of maternal histone messenger RNA in the sea urchin egg. *Proc. Nat. Acad. Sci., USA*, **79**, 5944–7.

Shur, B.D. and Hall, N.G. (1982a) Sperm surface galactyosyltransferase activities during *in vitro* capacitation. *J. Cell Biol.*, **95**, 567–73.

Shur, B.D. and Hall, N.G. (1982b) A role for mouse sperm surface galactosyltransferase in sperm binding to the egg zona pellucida. *J. Cell Biol.*, **95**, 574–9.

Simmel, E.B. and Karnofsky, D.A. (1961) Observations on the uptake of tritiated thymidine in the pronuclei of fertilized sand dollar embryos. *J. Biophys Biochem. Cytol.*, **10**, 59–65.

Singer, S.J. and Nicolson, G.L. (1972) The fluid mosaic model of the structure of cell membranes. *Science*, **175**, 720–31.

Skoblina, M. (1976) Role of karyoplasm in the emergence of capacity of egg cytoplasm to induce DNA synthesis in transplanted sperm nuclei. *J. Embryol, Exp. Morphol.*, **36**, 67–72.

Smith, L.D. and Ecker, R.E. (1970) Uterine suppression of biochemical and morphogenetic events in *Rana pipiens*. *Dev. Biol.*, **22**, 622–37.

Sorokin, S.P. (1968) Reconstructions of centriole formation and ciliogenesis in mammalian lungs. *J. Cell Sci.*, **3**, 207–30.

Spudich, A. and Spudich, J.A. (1979) Actin in triton-treated cortical preparations of unfertilized and fertilized sea urchin eggs. *J. Cell Biol.*, **82**, 212–26.

Stambaugh, R., Smith, M. and Faltas, S. (1975) An organized distribution of acrosomal proteinase in rabbit sperm acrosomes. *J. Exp. Zool.*, **193**, 119–22.

Stefanini, M., Oura, C. and Zamboni, L. (1969) Ultrastructure of fertilization in the mouse. II. Penetration of sperm into the ovum. *J. Submicrosc. Cytol.*, **1**, 1–23.

Steinhardt, R.A. and Epel, D. (1974) Activation of sea-urchin eggs by a calcium ionophore. *Proc. Natl. Acad. Sci., USA*, **71**, 1915–19.

Steinhardt, R.A., Epel, D., Carroll, E.J. and Yanagimachi, R. (1974) Is calcium ionophore a universal activator for unfertilized eggs? *Nature*, **252**, 41–3.

Steinhardt, R.A., Zucker, R. and Schatten, G. (1977) Intracellular calcium at fertilization in the sea urchin egg. *Dev. Biol.*, **58**, 185–96.

Steinman, R.M. (1968) An electron microscopic study of ciliogenesis in developing epidermis and trachea in the embryo of *Xenopus laevis*. *Amer. J. Anat.*, **122**, 19–56.

Storey, B.T., Lee, M.A., Muller, C. et al. (1984) Binding of mouse spermatozoa to the zona pellucidae of mouse eggs in cumulus: Evidence that the acrosomes remain substantially intact. *Biol. Reprod.*, **31**, 1119–28.

Summers, R.G. and Hylander, B.L. (1975) Species-specificity of acrosome reaction and primary gamete binding in echinoids. *Exp. Cell Res.*, **96**, 63–8.

Swenson, K.I., Farrell, K.M. and Ruderman, J.V. (1986) The clam embryo protein cyclin A induces entry into M phase and the resumption of meiosis in *Xenopus* oocytes. *Cell*, **47**, 861–70.

Szollosi, D. (1965) The fate of sperm middle piece mitochondria in the rat egg. *J. Exp. Zool.*, **159**, 367–78.

Szollosi, D. and Ris, H. (1961) Observations on sperm penetration in the rat. *J. Biophys. Biochem. Cytol.*, **10**, 275–83.

Szollosi, D., Calarco, P. and Donahue, R.P. (1972) Absence of centrioles in the first and second meiotic spindles of mouse oocytes. *J. Cell Sci.*, **11**, 521–41.

Tash, J.S. and Means, A.R. (1982) Regulation of protein phosphorylation and motility of sperm by cyclic adenosine monophosphate and calcium. *Biol. Reprod.*, **26**, 745–63.

Thadani, V.M. (1979) Injection of sperm heads into immature rat oocyte. *J. Exp. Zool.*, **210**, 161–8.

Thibault, C. and Gérard, M. (1973) Cytoplasmic and nuclear maturation of rabbit oocytes *in vitro*. *Ann. Biol. Anim. Bioch. Biophys.*, **13**, 145–56.

Tilney, L.G. (1975a) The role of actin in nonmuscle cell motility, in *Molecules and Cell Movement* (eds S. Inoué and R.E. Stephens), Raven Press, New York, pp. 339–88.

Tilney, L.G. (1975b) Actin filaments in the acrosomal reaction of *Limulus* sperm. *J. Cell Biol.*, **64**, 289–310.

Tilney, L.G. (1978) The polymerization of actin. V. A new organelle, the actomere, that initiates the assembly of actin filaments in *Thyone* sperm. *J. Cell Biol.*, **77**, 551–64.

Tilney, L.G. and Jaffe, L.A. (1980) Actin, microvilli and the fertilization cone of sea urchin eggs. *J. Cell Biol.*, **87**, 771–82.

Tilney, L.G., Kiehart, D.P., Sardet, C. and Tilney, M. (1978) Polymerization of actin. IV. Role of Ca^{2+} and H^+ in the assembly of actin and in membrane fusion in the acrosomal reaction of echinoderm sperm. *J. Cell Biol.*, **77**, 536–50.

Trounson, A.O., Willadsen, S.M. and Rowson, L.E.A. (1977) Fertilization and development capability of bovine follicular oocyte matured *in vitro* and *in vivo* and transferred to the oviducts of rabbits and cows. *J. Reprod. Fert.*, **51**, 321–7.

Turner, P.R., Sheetz, M.P. and Jaffe, L.A. (1984) Fertilization increases the polyphosphoinositide content of sea urchin eggs. *Nature*, **310**, 414–5.

Tyler, A. (1940) Sperm agglutination in the keyhole limpet, *Megathura crenulate*. *Biol. Bull.*, **78**, 159–78.

Uehara, T. and Yanagimachi, R. (1976) Microsurgical injection of spermatozoa into hamster eggs with subsequent transformation of sperm nuclei into male pronuclei. *Biol. Reprod.*, **15**, 467–70.

Uehara, T. and Yanagimachi, R. (1977) Behavior of nuclei of testicular, caput and cauda epididymal spermatozoa injected into hamster eggs. *Biol. Reprod.*, 16, 315–21.

Unsworth, B.R. and Kaulenas, M.S. (1975) Changes in ribosome-associated proteins during sea urchin development. *Differentiation*, 3, 21–27.

Usui, N. and Yanagimachi, R. (1976) Behavior of hamster sperm nuclei incorporated into eggs at various stages of maturation, fertilization and early development: The appearance and disappearance of factors involved in sperm chromatin decondensed in egg cytoplasm. *J. Ultrastruct. Res.*, 57, 276–88.

Vacquier, V.D. (1975) The isolation of intact cortical granules from sea urchin eggs: Calcium ions trigger granule discharge. *Dev. Biol.*, 43, 62–74.

Vacquier, V.D. (1979) The fertilizing capacity of sea urchin sperm rapidly decreases after induction of the acrosome reaction. *Dev. Growth Differ.*, 21, 61–69.

Vacquier, V.D. (1980) The adhesion of sperm to sea urchin eggs, in *The Cell Surface: Mediator of Developmental Processes* (eds S. Subtelny and N.K. Wessels), Academic Press, New York, pp. 151–68.

Vacquier, V.D. (1981) Dynamic changes of the egg cortex. *Dev. Biol.*, 84, 1–26.

Vacquier, V.D. and Moy, G.W. (1977) Isolation of bindin: The protein responsible for adhesion of sperm to sea urchin eggs. *Proc. Natl. Acad. Sci., USA*, 74, 2456–60.

Vacquier, V.D. and Moy, G.W. (1978) Macromolecules mediating sperm–egg recognization and adhesion during sea urchin fertilization, in *Cell Reproduction: Essays in Honor of Daniel Mazia* (eds E.R. Dirksen, D.M. Prescott and C.F. Fox), Academic Press, New York, pp. 379–89.

Vacquier, V.D. and Payne, J.E. (1973) Methods for quantitating sea urchin sperm–egg binding. *Exp. Cell Res.*, 82, 227–35.

Vacquier, V.D., Tegner, M.J. and Epel, D. (1973) Protease released from sea urchin eggs at fertilization alters the vitelline layer and aids in preventing polyspermy. *Exp. Cell Res.*, 80, 111–19.

Van Blerkom, J. (1977) Molecular approaches to the study of oocyte maturation and embryonic development, in *Immunobiology of the Gametes*, (eds. M. Edidin and M.H. Johnson), Cambridge University Press, Cambridge, pp. 187–202.

Van Blerkom, J. (1979) Molecular differentiation of the rabbit ovum. III. Fertilization-antonomous polypeptide synthesis. *Dev. Biol.*, 72, 188–94.

Van Blerkom, J. (1981) Structural relationships and post-translational modification of stage-specfic proteins synthesized during early preimplantation development in the mouse. *Proc. Natl. Acad. Sci., USA*, 78, 7629–33.

Van Blerkom, J. and Motta, P. (1979) *The Cellular Basis of Mammalian Reproduction.* Urban & Schwarzenberg, Baltimore.

Van Meel, F.C.M. and Pearson, P.L. (1979) Do human spermatozoa reactivate in the cytoplasm of somatic cells? *J. Cell. Sci.*, 35, 105–22.

Veron, M. and Shapiro, B.M. (1977) Binding of concanavalin A to the surface of sea urchin eggs and its alteration upon fertilization. *J. Biol. Chem.*, 252, 1286–92.

Veron, M., Foerder, C., Eddy, E.M. and Shapiro, B.M. (1977) Sequential biochemical and morphological events during assembly of the fertilization membrane of the sea urchin. *Cell*, 10, 321–8.

Wassarman, P.M., Josefowicz, W.J. and Letourneau, G.E. (1976) Meiotic maturation of mouse oocytes *in vitro*: Inhibition of maturation at specific stages of nuclear progression. *J. Cell Sci.*, **22**, 531–45.

Wassarman, P.M., Bleil, J.D., Cascio, S.M. et al. (1981) Programming of gene expression during mammalin oogenesis, in *Bioregulators of Reproduction* (eds G. Jagiello and H.J. Vogel), Academic Press, New York, pp. 119–50.

Wasserman, W.J. and Masui, Y. (1975) Effects of cycloheximide on a cytoplasmic factor initiating meiotic maturation in *Xenopus* oocytes. *Exp. Cell Res.*, **91**, 381–8.

Wasserman, W.J. and Smith, L.D. (1978) The cyclic behaviour of a cytoplasmic factor controlling nuclear membrane breakdown. *J. Cell Biol,*, **78**, R15–R22.

Weisenberg, R.C. and Rosenfeld, A.C. (1975) *In vitro* polymerization of microtubules into asters and spindles in homogenates of surf clam eggs. *J. Cell Biol.*, **64**, 146–58.

Whitaker, M.J. and Steinhardt, R.A. (1981) The relation between the increase in reduced nicotinamide nucleotides and the initiation of DNA synthesis in sea urchin eggs. *Cell*, **25**, 95–103.

Whitaker, M.J. and Steinhardt, R.A. (1982) Ionic regulation of egg activation. *Quart. Rev. Biophys.*, **15**, 593–666.

Whitaker, M.J. and Steinhardt, R.A. (1983) Evidence in support of the hypothesis of an electrically mediated fast block to polyspermy in sea urchin eggs. *Dev. Biol.*, **95**, 244–8.

Whitaker, M.J. and Irvine, R.F. (1984) Inositol 1,4,5 trisphosphate microinjection activates sea urchin eggs. *Nature*, **312**, 636–9.

Wiesel, S. and Schultz, G.A. (1981) Factors which may affect removal of protamine from sperm DNA during fertilization in the rabbit. *Gam. Res,*, **4**, 25–34.

Wilson, E.B. (1925) *The Cell in Development and Heredity*, Macmillan, New York.

Winkler, M.M., Steinhardt, R.A., Grainger, J.L. and Minning, L. (1980) Dual ionic controls for the activation of protein synthesis at fertilization. *Nature*, **287**, 558–60.

Wolf, D.P. and Hamada, M. (1977) Induction of zonal and egg plasma membrane blocks to sperm penetration in mouse eggs with cortical granule exudate. *Biol. Reprod.*, **17**, 350–4.

Wolf, D.P., Nicosia, S.V. and Hamada, M. (1979) Premature cortical granule loss does not prevent sperm penetration of mouse eggs. *Dev. Biol.*, **71**, 22–32.

Wolf, D.E. and Ziomek, C.A. (1983) Regionalization and lateral diffusion of membrane proteins in unfertilized and fertilized mouse eggs. *J. Cell Biol.*, **96**, 1786–90.

Wolf, D.E., Edidin, M. and Handyside A.H. (1981) Changes in the organization of the mouse egg plasma membrane upon fertilization and first cleavage. Indications from the lateral diffusion rates of fluorescent lipid analogs. *Dev. Biol.*, **85**, 195–8.

Wolf, D.E., Kinsey, W., Lennarz, W. and Edidin, M. (1981) Changes in the organization of the sea urchin egg plasma membrane upon fertilization: Indications from the lateral diffusion rates of lipid-soluble fluorescent dyes. *Dev. Biol.*, **81**, 133–8.

Wolf, R. (1978) The cytaster, a colchicine-sensitive migration organelle of cleavage nuclei in an insect egg. *Dev. Biol.*, **62**, 464–72.

Wu, J.T. and Chang, M.C. (1973) Reciprocal fertilization between the ferret and short-tailed weasel with special reference to the development of ferret eggs fertilized by weasel sperm. *J. Exp. Zool.*, **183**, 281–90.

Wu, M. and Gerhardt, J.C. (1980) Partial purification and characterization of the maturation-promoting factor from eggs of *Xenopus laevis*. *Dev. Biol.*, **79**, 465–77.

Wyrick, R.E., Nishihara, T. and Hedrick, J.L. (1974) Agglutination of jelly coat and cortical granule components and the block to polyspermy in the amphibian *Xenopus laevis*. *Proc. Natl. Acad. Sci., USA*, **71**, 2067–71.

Yamamoto, K. and Yoneda, M. (1983) Cytoplasmic cycle in meiotic division of starfish oocytes. *Dev. Biol.*, **96**, 166–72.

Yanagimachi, R. (1981) Mechanisms of fertilization in mammals, in *Fertilization and Embryonic Development In Vitro* (eds L. Mastroianni and D.J. Biggers), Plenum Press, New York, pp. 81–182.

Yanagimachi, R. and Noda, Y.D. (1970a) Ultrastructural changes in the hamster sperm head during fertilization. *J. Ultrastruct. Res.*, **31**, 465–85.

Yanagimachi, R. and Noda, Y.D. (1970b) Electron microscopic studies of sperm incorporation into the golden hamster egg. *Amer. J. Anat.*, **128**, 429–62.

Yanagimachi, R. and Nicolson, G. (1976) Lectin-binding properties of hamster egg zona pellucida and plasma membrane during maturation and preimplantation development. *Exp. Cell Res.*, **100**, 249–57.

Yanagimachi, R., Nicolson, G.L., Noda, Y.D. and Fujimoto, M. (1973) Electron microscopic observations of the distribution of acidic anionic residues on hamster spermatozoa and eggs before and during fertilization. *J. Ultrastruct. Res.*, **43**, 344–53.

Yasumasu, I., Tazawa, E. and Fujiwara, A. (1975) Glycolysis in the eggs of the echiuroid, *Urechis unicinctus* and the oyster, *Crassostrea gigas*. Rate-limiting steps and activation at fertilization. *Exp. Cell Res.*, **93**, 166–74.

Yoneda, M., Ikeda, M. and Washitani, S. (1978) Periodic change in the tension at the surface of activated non-nucleate fragments of sea urchin eggs. *Dev. Growth Differ.*, **20**, 329–36.

Young, E.M. and Raff, R.A. (1979) Messenger ribonucleoprotein particles in developing sea urchin embryos. *Dev. Biol.*, **72**, 24–40.

Young, R.J. (1979) Rabbit sperm chromatin is decondensed by a thiol-induced proteolytic activity not endogenous to its nucleus. *Biol. Reprod.*, **20**, 1001–04.

Young, R.J. and Sweeney, K. (1978) Mammalian ova and one-cell embryos do not incorporate phosphate into nucleic acids. *Eur. J. Biochem.*, **91**, 111–17.

Yu, S.F. and Wolf, D.P. (1981) Polyspermic mouse eggs can dispose of supernumerary sperm. *Dev. Biol.*, **82**, 203–10.

Zamboni, L. (1971) *Fine Morphology of Mammalian Fertilization*, Harper and Row, New York.

Zamboni, L., Mishell, D.R., Bell, J.H. and Baca, M. (1966) Fine structure of the human ovum in the pronuclear stage. *J. Cell Biol.*, **30**, 579–600.

Zimmerman, A.M. and Forer, A. (1981) *Mitosis/Cytokinesis*, Academic Press, New York.

Zimmerman, A.M. and Zimmerman, S. (1967) Action of colcemid in sea urchin eggs. *J. Cell Biol.*, **34**, 483–8.

Zirkin, B.R. and Chang, T.S.K. (1977) Involvement of endogenous proteolytic activity in thiol-induced release of DNA template restrictions in rabbit sperm nuclei. *Biol. Reprod.*, **17**, 131–7.

Zirkin, B.R., Chang, T.S.K. and Heaps, J. (1980) Involvement of an acrosin-like proteinase in the sulfhydryl-induced degradation of rabbit sperm nuclear protamine. *J. Cell Biol.*, **85**, 116–21.

Zucker, R.S., Steinhardt, R.A. and Winkler, M.M. (1978) Intracellular calcium release and the mechanisms of parthenogenetic activation of the sea urchin egg. *Dev. Biol.*, **65**, 285–95.

Index